Arms and the Men

BY THE SAME AUTHOR

The Defence of the United Kingdom
The Battle of Britain
The Battle of the V-Weapons 1944–45
A Short History of the Second World War
The War in the Far East 1941–45

BASIL COLLIER

Arms and the Men

The Arms Trade and Governments

HAMISH HAMILTON
LONDON

First published in Great Britain 1980
by Hamish Hamilton Ltd
Garden House 57–59 Long Acre London WC2E 9JZ

British Library Cataloguing in Publication Data

Collier, Basil
Arms and the men.
1. Munitions – History.
I. Title
338.4′7′62340904 HD9743.A2
ISBN 0–241–10308–8

Printed in Great Britain by
Ebenezer Baylis & Son Limited
The Trinity Press, Worcester, and London

When the sands are all dry, he is gay as a lark,
And will talk in contemptuous tones of the Shark:
But, when the tide rises and sharks are around,
His voice has a timid and tremulous sound.

—Lewis Carroll *'Tis the Voice of the Lobster*

Now thrive the armourers, and Honour's thought
Reigns solely in the breast of every man.
They sell the pasture now, to buy the horse.

—William Shakespeare, *Henry V*, Act II, Prologue

Contents

Illustrations

Between pages 148 *and* 149

Photo credits

Plates 1a, 1c, 1d, 2a, 2b, 3a, 3b, 3c, 5a, 6a, 7a *Radio Times Hulton Picture Library*;

Plates 2c, 4b, 5c, 8b, 8c *Peter Newark's Historical Pictures*;

Plates 1b, 5b *Peter Newark's Western Americana*;

Plate 4a by kind permission of *Vickers Limited*;

Plate 6b *Keystone Press Agency*;

Plate 7b by permission of the Trustees of the Imperial War Museum;

Plate 8a *Popperfoto*;

Page 188 *Peter Newark's Western Americana*;

Page 237 *Peter Newark's Western Americana*

Author's Preface

The purpose of this book is to give some account of the genesis and development of the international trade in arms, the relations that have subsisted at various times between arms manufacturers and governments, and the means by which nations have equipped themselves for war. The first of these themes forms part of the subject matter of early chapters, devoted largely to the discussion of technical innovations and social trends whose effect was to shift the onus of providing weapons of war from the individual warrior to the state. The second and third themes recur throughout the book, but are treated chiefly in the context of problems of procurement and supply encountered before and after the outbreak of hostilities by some of the leading belligerents in two world wars. A chapter and more are given to the remarkable man we knew as Basil Zaharoff. Finally, reference is made in the last chapter to current problems, and to economic pressures which have led since the end of World War II to direct participation by some governments in the sale of arms to foreign countries.

With the exception of a few laudatory references by romantic versifiers to armourers and swordsmiths, seldom precisely located in time or space, arms manufacturers have not received much favourable notice from imaginative writers, or indeed from writers of any category. (Shaw's Undershaft, for example, is not, and is not meant to be, an attractive figure.) The reason, I think, is not so much that arms manufacturers tend not to possess characteristics likely to appeal to literary men—Trevor Dawson and Charles Craven did possess such characteristics, and Meade Falkner was himself a novelist—as that few writers have gone out of their way to question the assumption that they don't and can't possess them.

Nor have they, on the whole, been much esteemed by the

general public, at any rate in peacetime. Like the Tommy Atkins of Kipling's poem they have sometimes received unaccustomed deference when the guns began to shoot; but the goodwill engendered by danger has seldom lasted long. By the last year of World War I there was a fairly widespread tendency to lump the heads of the great armament firms with generals and statesmen as enemies of the common and even the uncommon man.

Between the wars arms manufacturers were freely accused, as a class, of trying to increase the demand for their wares by spreading false reports and fomenting war scares, of seeking by improper means to influence public opinion in favour of armament programmes, of conspiring with each other to raise prices. I show that these charges were based not, as a rule, on any factual evidence which those who made them claimed to be able to produce, but on the erroneous belief that their truth had already been established.

For advice based on an unsurpassed knowledge of problems of procurement and supply in time of war I owe the late Lord Chandos (better known as Oliver Lyttelton) a debt which only his untimely death prevented from increasing. I am indebted to the directors of Vickers Limited for opening their archives to me and helping my researches in other ways, and to Lord Glenkinglas for giving me a clearer insight than I could otherwise have hoped to have into the character of his grandfather, Sir Andrew Noble. To them and others I tender my warm thanks.

B.C.

March, 1979

ONE

Merchants of Death?

The procurement of arms is one of the most important and most exacting tasks performed by governments. It is a responsibility which no government, irrespective of its political complexion, can escape. As long as lethal weapons are deemed necessary for internal and external security, statesmen and officials feel bound to provide them. They know that, at any rate in peacetime, they can expect from the great mass of the general public neither gratitude for their efforts nor a sympathetic understanding of their problems. About the most a Minister of Defence and his co-adjutors can hope for is that the veil of secrecy which tends to shroud their activities will never be lifted by accusations of corruption or failure to provide enough of the right weapons.

Still less can manufacturers and suppliers of arms expect more than a handful of their fellow-citizens to be grateful for their contributions to national defence and the maintenance of law and order. The attitude of most of us to arms manufacturers resembles that attributed by ethnologists to the Masai of East Africa. The Masai are a warlike people, much given within living memory to the use of swords and spears in feuds and cattle-raiding expeditions which formed a great part of the business of their lives. But a sharp distinction was, and perhaps still is, drawn between the wielding and the making of these weapons. Masai ironsmiths, because some of their products are potential instruments of bloodshed, used to be eyed askance, and probably still are eyed askance by the more conservative members of the community. They are, or at any rate were until quite recently, allowed to marry only amongst themselves, and ethnologists tell us that 'even extra-marital relations with their women' were 'considered dangerous'.*

In the Western world, we, too, have had our families of

* Forde, *Habitat, Economy and Society*, 298.

hereditary ironsmiths who provided us with lethal weapons. Schneider, Krupp and Vickers—the three best-known arms manufacturers of the steam age—all began as iron or steel firms owned and administered by partners or sole proprietors who brought up their sons to follow in their footsteps. The American Andrew Carnegie made a vast fortune as a steel manufacturer. He spent millions of dollars on attempts to promote universal peace. At one time he disclaimed any intention of ever making armour plate for American or foreign warships, but eventually his firm earned huge profits by doing so.

That, however, was yesterday. Today we call providers of the means of making war or suppressing insurrections not arms manufacturers but manufacturers of defence equipment. This term covers a vast array of products, ranging from bulldozers and intercontinental bombers to fire-extinguishers, uniforms and buttons. By that definition at least five hundred firms in Britain alone, and perhaps three times as many in the United States, can be classed as whole-time or part-time producers of instruments of war. At the same time, the more sophisticated weapons and weapons-systems have become so complex that their design and development, although not necessarily their manufacture, tends to be monopolized by a relatively small number of government-controlled or free-enterprise organizations with huge capital resources furnished largely by ordinary members of the public in their capacity as investors or taxpayers. The wealth of modern highly-industrialized capitalist societies is so widely distributed, relations between government departments and financial and commercial organizations have become so complex, that probably no British or American citizen who pays national or federal taxes, contributes to a pension fund, holds a life insurance policy, belongs to a trade union or entrusts his money to a bank can claim with complete confidence that he has no stake in the private—let alone the public—sector of the arms trade.

Furthermore, in modern capitalist societies the production of defence material is sustained not only by the involuntary participation of millions of taxpayers and contributors to funds administered by banks, insurance companies and investment trusts, but also by many thousands of willing investors in corporations or limited liability companies large and small. Nearly half the ordinary stock in one of the world's best-known armament firms is held by investors with an average holding of well under five hundred pounds, or say a thousand dollars. How do these and other shareholders in enterprises associated with the manufacture of defence material reconcile their financial interest in the multiplication of instruments of war with the humanitarian sentiments which the spirit of the age has made almost obligatory in the Western world?

Many, of course, don't try to do so. If pressed, they might say that weapons of war are a necessary evil, that a lover of mankind is no more bound to dissociate himself from their manufacture and distribution than a lover of animals is bound to be a vegetarian. Others find comfort in the reflection that few companies or corporations large enough to offer stocks or shares for public subscription concern themselves exclusively with the production of arms. Masai ironsmiths make cowbells as well as swords and spearheads. Most of the big American airspace corporations offer civil as well as naval or military aircraft. Among British firms, Vickers Limited and its subsidiary and associated companies are producers not only of escort and patrol vessels, nuclear-powered submarines, naval guns and gun-mountings, ballistic missiles, integrated fire-control systems and guided-missile launchers but also of medical and optical equipment, office equipment and supplies, printing machinery, malting and bottling plant, lithographic plates, submersible vessels for wreck-finding and other commercial and scientific purposes, and a wide range of general engineering products. Ferranti Limited are suppliers of air defence systems, digital fire-control systems and weapon delivery systems, but they also supply air traffic control systems and flight instruments for all types of aircraft. Alfried Krupp, in the days when he was called the Cannon King, derived much of his large income from the sale of seamless steel tyres to the Prussian State Railways.

That is one way of looking at the matter. But it would be unrealistic to suppose that more than a very few manufacturers of defence material have ever regarded themselves as people for whom excuses must be made. The armament kings of the late nineteenth and early twentieth centuries may not have been popular. They were reviled at times by critics of the capitalist system and by pacifists. Satisfied customers and fellow-industrialists saw them in a different light. The salaried company directors and senior civil servants on whom their mantle has fallen see no reason to apologize for predecessors whom they regard as having conferred substantial benefits on their fellow-citizens in peace and war. 'People praise the Spitfire pilots of the Second World War,' a spokesman for the arms industry told the author of this book. 'We made the Spitfire, and they call us merchants of death.'

That argument would be almost impossible to counter if manufacturers of defence material supplied only the armed forces of their own countries. But that is not the case. Since the middle of the nineteenth century all the leading armament firms have maintained their productive capacity in peacetime by selling arms abroad, or at any rate have tried to do so. In the present state of the world they are not merely allowed but actively encouraged by their governments to seek foreign business. It is not hard to see why. Exports of defence material not only earn profits and create jobs, but help governments to redress trade balances, maintain reserves of gold and foreign currencies, and meet the cost of equipping their own armed forces. Communists as well as capitalist statesmen are alive to these benefits. Soviet Russia and her satellite states—especially Czechoslovakia—are eager exporters of arms, not merely for ideological and strategic but also for economic reasons. The British government employs a staff of experts for the twofold purpose of promoting exports of defence equipment by commercial firms and selling to foreign customers the products of its own arms factories. France and the United States, to say nothing of the Soviet Union, likewise employ officials who are, in effect, government arms salesmen paid from public funds.

Many people find this shocking. That arms manufacturers are allowed, let alone encouraged, to sell their products to foreign governments seems to them immoral. Their attitude may not be logically defensible. Nevertheless there is no doubt that much of the odium which attaches to the arms business in the estimation of ordinary citizens stems from its international character.

However, to give the impression that arms manufacturers have never been suspected of anything worse than selling their goods to foreigners would be highly misleading. They have repeatedly, and sometimes justly, been accused of bribing, or attempting to bribe, officials in their own or in foreign countries. As a class—though seldom individually—they have also been charged with fomenting war scares, disseminating false reports about the plans and intentions of potentially hostile powers, seeking to influence public opinion in favour of rearmament programmes, and conspiring with each other to raise prices and to boost the demand for their wares by fomenting international rivalries.

Arms manufacturers have also been accused of making huge profits for themselves or their shareholders by exploiting their workpeople. That accusation may have had some force in the days when workpeople were more easily exploited than they are now. Even so, the record of the leading armament firms as employers of labour is not noticeably worse than that of other large concerns. In some respects it is better. Schneider et Compagnie were pioneers of such social benefits as housing schemes and educational facilities for employees and their families, pensions for the superannuated and free medical and pharmaceutical services. Alfried Krupp and his successors prided themselves on their paternalism at a time when paternalism was not frowned upon by trade unionists but was still regarded as a merit. The leading British armament firms had their share of labour troubles before, during and immediately after the First World War, but by the middle of the 1920s relations between employers and employed at the Vickers works at Barrow were so good that the General Manager of the Shipbuilding Division received a patient hearing when he called his workpeople together on the eve of a General Election and addressed them on behalf of the Conservative candidate.

The charge of corruption is in a different category from the other reproaches commonly levelled against arms manufacturers. No one doubts that massive sums have on various occasions been spent on bribery or attempted bribery both by armament firms and by other large concerns which do business with governments. The origin of the remaining charges can be traced largely, if not wholly, to efforts made from time to time by critics of the capitalist system to establish the complicity of rich and powerful firms in attempts to stimulate the demand for their products by improper means. The Russian Revolution of 1917 was followed by a blanket charge against Western capitalists of seeking to prolong the war in order to increase their profits. Although this argument attracted some attention in intellectual circles at the time, it was not until the capitalist world was shaken by the after-effects of the Wall

Street crash of 1929 that widespread and persistent attempts were made to convert large numbers of ordinary men and women to the belief that arms manufacturers must be regarded as enemies of society. A number of books which contained bitter attacks on the arms industry then appeared in various countries. One of the most influential was *The Bloody Traffic*, by Fenner Brockway (now Lord Brockway).* At that time Fenner Brockway was an ardent exponent of capitalist errors. He accused arms manufacturers, as a class, not only of bribing officials but also of conspiring with each other to boost sales and force up prices, using their influence with the Press and in financial circles to further their own interests,[1] and selling arms to hostile or potentially hostile countries while pretending to be patriotic. He backed his case by asserting that a Temporary Mixed Commission set up by the League of Nations to make proposals for the reduction of armaments reported that armament firms had:

1 Been active in fomenting war scares and in persuading their own countries to adopt warlike policies and increase their armaments;
2 Attempted to bribe government officials at home and abroad;
3 Disseminated false reports concerning the military and naval programmes of various countries, in order to stimulate expenditure on armaments;
4 Sought to influence public opinion through the control of newspapers in their own and foreign countries;
5 Organized international armament rings whose effect was to accentuate the arms race by the playing-off of one country against another;
6 Organized international armament trusts which increased the cost of armaments sold to governments.

This assertion has been repeated by a number of authors. It has been made as recently as 1972 in a book published under the auspices of the International Institute for Strategic Studies.†

But Brockway and those who followed him were wrong. No one

* Fenner Brockway, *The Bloody Traffic*. London: Gollancz, 1933.

† John Stanley and Maurice Pearton. *The International Trade in Arms*. London: Chatto and Windus for the International Institute for Strategic Studies, 1972.

questions their good faith or their sincerity. The fact remains that the Temporary Mixed Commission appointed by the League of Nations in 1921 did *not* report that armament firms had indulged in the practices listed. It merely stated in its report that these were the headings under which, in general, objections that had been raised to the 'untrammelled' private manufacture of armaments could be grouped.* One member of the Temporary Mixed Commission is known to have interpreted the report as an endorsement of the objections, but there is ample evidence that it was not meant to bear that interpretation. The sub-committee which dealt with the private manufacture of armaments and drafted the relevant part of the report did not investigate the objections. It merely discussed the *prima facie* case against uncontrolled private manufacture in the light of each member's knowledge of what was being said about the arms trade by its critics. The member of the Temporary Mixed Commission who believed that the commission had endorsed the objections was not a member of the sub-committee and took no part in these discussions.

So the report of the Temporary Mixed Commission does not throw much light, or indeed any light, on whether arms manufacturers were or were not addicted to the goings-on imputed to them in 1921. Anyone concerned to substantiate the allegations made then and repeated later must look elsewhere for his powder and shot. If he searches long enough he will find evidence in reports of departmental investigations or criminal proceedings that officials in various countries have sometimes been offered bribes and on rare occasions have been prosecuted for accepting them. He may also find evidence of the payment of massive entertainment allowances or special commissions to foreign agents or emissaries. In view of the notorious difficulty of transacting business in certain countries without bribery, and in the light of admissions made from time to time by persons connected with the arms trade, he may feel tolerably sure that some of this money found its way into the pockets or numbered bank accounts of corrupt officials.

To prove that in any particular instance a bribe was given and received is another matter. Most agents of big firms engaged in

* *Report of the Temporary Mixed Commission on Armaments*. League of Nations Document No. A 81 of 1921.

negotiations with foreign governments relied at one time largely on freelance sub-agents and go-betweens for background knowledge of the political and social scene, introductions to ministers and officials, and tip-offs about their standing with politicians and the activities of competitors. To some extent they still do so, even though their own governments may now be in a position to give them all the introductions they need and tell them most of what they want to know. A payment made to an intermediary for effecting an introduction or arranging a dinner party is not a bribe. Whether a payment made for information received is corrupt depends upon how the information was obtained. If, as often happens, an agent pays an informant for an invented story, or for passing on mere gossip picked up in ante-rooms and lobbies, he may have wasted his firm's money, but he has not bribed anyone.

Moreover, even what may seem to a reader of newspaper reports incontrovertible evidence of bribery cannot always be taken at its face value. Let us suppose that a firm with world-wide ramifications has tendered for a multi-million-dollar defence contract in Ruritania. A Ruritanian known to have contacts in high places approaches the firm's agent, tells him that the government has come within a hair's breadth of accepting his tender, but adds that an official who must sanction the deal will not do so unless the firm pays $50,000 into a Swiss bank account which cannot be traced to him. A great deal is at stake for the agent and his principal. He makes arrangements with the firm's representative in Switzerland for the money to be paid and to be entered in the accounts under some innocuous heading. The firm is awarded the contract, because that was going to happen anyway. The agent is delighted. The astute Ruritanian, whose sole asset was the knowledge that the official had already made up his mind to sanction the deal, becomes the richer by $50,000 in addition to a fee or honorarium paid by the grateful firm. Moreover, he acquires a great reputation as a fixer. If the transaction afterwards comes to light as the result of some mishap or indiscretion, the official may find it almost impossible to prove that he was not bribed, although in fact he never received a penny. At least three people—the agent, the firm's representative in Switzerland and the supposed go-between—are prepared to say that the money was paid into a bank account understood to be his, and the bank manager is forbidden by Swiss law to say what happened to it.

Allegations that arms manufacturers have tried to boost sales by giving governments false information about preparations for war in potentially hostile countries can seldom be related to individual firms and specific cases. Even when they can, the evidence tends not to amount to very much. The highly successful arms salesman, Basil Zaharoff, is supposed about the middle of the 1880s to have sold three submarines on behalf of his employer, Thorsten Nordenfelt, by telling the Greeks that they needed a submarine to defend themselves against their hereditary enemies the Turks, and the Turks that they needed two submarines because the Greeks had bought one. We know that the sales were made and that Zaharoff was Nordenfelt's agent or correspondent in the Near East at the time. Even if the rest of the story is true, there is nothing in it that would justify a charge against Zaharoff or Nordenfelt of fomenting a war-scare or spreading a false report. Whether the Greeks needed a submarine could only be a matter of opinion, and the information supposed to have been given to the Turks was accurate.

Again, Zaharoff is alleged to have given a wilfully misleading account of the result of comparative trials of the Nordenfelt quick-firing gun and the Maxim machine-gun, held near Vienna on 7 May 1888. The Maxim proved an easy winner, but Zaharoff is said to have declared that it was the Nordenfelt, not the Maxim, that carried all before it. He may well have inspired a laudatory account of the Nordenfelt that appeared in the Austrian press, but it seems very unlikely that he tried to deceive the Austrian military authorities. In the first place, they could see for themselves that the Maxim was more accurate than the Nordenfelt and needed a smaller crew. Secondly, the story depicts Zaharoff as urging them to buy the well-tried Nordenfelt in preference to a new weapon which gave good results when fired by its inventor but had yet to prove its worth in other hands. But he can scarcely have said this, for he and they knew very well that the Maxim was not new. It was new to the Austrians in the sense that they had not yet adopted it, but was far from being an untried weapon. The original, rifle-calibre version had been in production since 1884, a 37-millimetre model which supplemented but did not replace it since 1885. The Austrian General Staff must have been well aware that by 1888 large numbers of Maxim guns had been sold and had given satisfaction in the hands of service users.

A further reason for regarding the story with suspicion is that, when the trials were held, the Maxim and Nordenfelt companies were on the point of joining forces. Zaharoff was a strong candidate for the post of foreign agent to the new company, and was in fact appointed. He was very well informed about the arms trade. It is hard to believe that he knew nothing of the impending merger and, in order to gain a doubtful benefit, would have risked antagonizing the directors of the Maxim company by telling lies about their product. No doubt, as long as the Nordenfelt company was his employer, he made out the best case he could for the Nordenfelt gun, but there is a big difference between doing that and giving a blatantly false account of trials witnessed by representatives of both companies.

Many other tales are told to illustrate the alleged addiction of arms manufacturers to dubious means of recommending their wares. They tend to be vague or, if specific, to relate to the activities of one man, Zaharoff. As we shall see in a later chapter, Zaharoff was shrewd and ambitious, seems to have enjoyed surrounding himself with an aura of mystery, and had some reason to shun the limelight. If a man has something to hide, one way of hiding it is for him to tell stories about himself which are more likely to raise a laugh than to be taken seriously. It seems probable that, just as many stories about the alleged meanness of the frugal but hospitable Scots are told by Scotsmen, a good many of the stories which depict Zaharoff as stealing a march on his rivals originated with Zaharoff himself. It will be noticed that, while he is sometimes represented in these anecdotes as more than a little unscrupulous, he is almost invariably represented as exceptionally quick-witted and resourceful. That might well be the reputation which a super-salesman who lived by selling expensive goods in a highly competitive market would wish to achieve.

There is another circumstance which suggests that Zaharoff may have played an active part in creating his own legend. A typical Zaharoff story depicts him as calling on a foreign statesman from whom he hopes to obtain a large order. The statesman says the order has already gone to another firm. 'I am sure,' says Zaharoff, 'that Your Excellency would not decide so important a matter except on mature consideration. You will wish to give it further thought. I will call tomorrow, Thursday, for your answer.' 'You would be wasting your time,' says the statesman, 'and anyway tomorrow isn't Thursday. It is Tuesday.' 'Let us have a little bet on it,' says Zaharoff. 'I will bet Your Excellency 10,000 francs that tomorrow is Thursday.' And he receives the order.

This is a good story, but is it true? If it is, then it must have been told in the first instance by Zaharoff, for the statesman would not have told it and no one else was present. If the story isn't true, almost anyone might have invented it, but few people would have been more likely to do so than Zaharoff. And if the story was a fabrication, then of course it is not evidence of the kind of thing that happened in the arms trade in the bad old days, but only evidence of the kind of thing that could be thought to happen.

It may, perhaps, be significant that Zaharoff is the central figure in an overwhelmingly high proportion of the stories told in the past, and still told today, to illustrate the wiles of arms manufacturers. Apart from the fact that he was not an arms manufacturer but a salesman, a negotiator and an expert on markets, one might expect that, if the stories had any evidential value, other leading figures in the arms trade would sometimes share the centre of the stage with him. Yet they seldom do. Stuart Rendel, afterwards 1st Baron Rendel, began selling arms abroad for Sir W. G. Armstrong and Company long before Zaharoff entered the service of the relatively obscure Nordenfelt. For some years he was in effect, if not in name, the company's London manager, with a special responsibility for foreign business. He was immensely successful. Yet for every anecdote in which Stuart Rendel figures, Zaharoff figures in a dozen. Is this solely because the sort of thing that happened to Zaharoff scarcely ever happened to Stuart Rendel? Or could one reason be that Rendel, a barrister with political ambitions, was not the kind of man who enjoys praising himself with faint damns, and that Zaharoff was such a man?

The fact remains that the 1st Sub-Committee of the Temporary Mixed Commission on Armaments made a fair assessment of the state of public opinion when in 1921 it drew up its list of objections commonly made to the unrestricted private manufacture of weapons of war. Armament firms *were* widely suspected, and some twelve years later were to be vehemently accused, not only of having in the past bribed officials, fomented war scares, spread false reports, conspired with each other to raise prices and used their supposed influence with the press and in governmental and financial circles to foster warlike policies, but also of seeking to perpetuate these abuses. That is understandable. In 1918 the armistice between Germany and the Allied and Associated Powers had been hailed with delirious joy as the prelude to an era of universal peace. Men weary of bloodshed, discomfort and privation were bitterly disappointed when the framing of the peace treaties was accompanied or followed by unedifying squabbles in Central and Eastern Europe, armed insurrections in Ireland, Morocco and the Middle East, sanguinary conflicts between Russians of the Right and Left, war between Russians and Poles, violence or threats of violence in India and Egypt, and a long-drawn struggle between Greeks and Turks in Asia Minor. These and other dire happenings must, they felt, have been contrived or assisted by men whose interests were bound up with slaughter and destruction. Who were more likely culprits than arms manufacturers?

This reasoning was not altogether sound. There were quite a number of people in Europe and elsewhere in the post-war era who had not only stronger motives than arms manufacturers for fostering real or supposed conflicts of national or ideological interest, but also better opportunities of doing so. Governments, not arms manufacturers, decide whether wars should be launched or dissident minorities in foreign countries be encouraged to take up arms. That is certain. To what extent, if at all, they have been influenced on a given occasion by pressure exerted by interested parties can only be a matter of conjecture. Statesmen who declare war or foment a rebellion because someone who hopes to profit by the transaction has urged or bribed them to do so, do not publish their reasons or record them in the national archives, so anyone who claims to know their motives is either guessing or has relied on imputations or hearsay evidence which may be either true or false.

Even so, no one will deny that the policies of the leading European powers for some years after the First World War reflected a lively fear of Communism and an anxious concern for the future of capitalist societies. When the armistice with Germany was signed, the Allied governments controlled large reserves of weapons and ammunition accumulated in the belief that fighting on the Western Front was not likely to end before the summer of 1919. While these lasted, they were well placed to back participants in relatively minor wars without courting electoral disfavour by ordering large numbers of new, expensive weapons. With the possible exception of Schneider, the leading European armament firms did not derive much financial advantage from the internecine conflicts of the early post-war period. Nor could they expect to do so. British arms manufacturers, in particular, knew that the signing of the armistice with Germany meant that many years might elapse before they received further large orders from the government for guns and shells. Fomenting conflicts between Poles and Russians, Greeks and Turks, and Communists and anti-Communists might or might not serve the interests of governments concerned with problems of national survival and with far-reaching political and economic issues. It could not save large wartime producers of arms whose fortunes were bound up with the welfare of heavy industry from having to recast their production schedules.

It may be asked what in fact the leading armament firms were doing in 1921 if they were not engaged in stirring up fresh troubles in Eastern Europe and the Near and Middle East. We shall see in a later chapter that some of them were giving part of their attention to attempts to find new markets for arms and the formation or revival of armament undertakings in foreign countries. But all of them were primarily concerned with the problem of surviving an unforeseen slump with investment and production programmes geared to the expectation that the return of peace would bring a steady and prolonged demand for products which had nothing to do with war. Gustav Krupp asked his co-directors within a few weeks of the armistice to agree that the Krupp factories should re-tool for the production of agricultural and textile machinery, dredgers, crankshafts for the engines of motor-cars and lorries, and a wide range of light engineering products and consumer goods. Since that time the firm had, however, received welcome orders for locomotives from the Prussian State Railways and the Soviet government. Vickers had practically no orders for armaments on their books at the end of 1921 or early in 1922, apart from one from Japan for armour plate. Their plan to meet post-war conditions was founded on a massive investment in the electrical industry and hopes of a shipbuilding boom and a ready market for railway material, forgings and stampings, boilers, turbines, reciprocating steam engines, diesel and gas engines, industrial machinery, sporting guns, sewing machines and motor-cars. Armstrongs, too, invested hugely after the armistice in undertakings not connected with the arms trade.

The Schneider company was not in quite the same position as Krupp, Vickers or Armstrongs, because France retained her large standing army after the war, consolidated her hold on Syria by force of arms, and used her diplomatic and financial resources to establish strong ties with Poland, Rumania and the Austrian succession states. Her military and colonial policies tended to maintain a fairly brisk demand for Schneider-made weapons, even when other nations were disarming. But Schneider had such a big stake in mining and metallurgical undertakings at home and abroad and in French industry as a whole, that attempts to interpret the company's policies and actions purely in terms of its armament interests, although often made, are bound to be misleading.

In any case there does not seem to be any very secure ground for the popular belief that French industrialists in general, and those who controlled the Schneider organization in particular, possessed between the First and Second World Wars some mysterious power which enabled them to bend governments to their will. French industrialists were no more able to prevent the government of the day from accepting the restrictions on naval shipbuilding imposed by the Naval Treaty of Washington than American, British, Italian and Japanese industrialists were able to prevent their governments from doing so. France also accepted the London Naval Treaty of 1930. On both occasions armament firms joined naval experts in voicing ineffective protests. Some of them were afterwards reproached with trying to defeat attempts to bring about disarmament agreements. Yet there would seem, on the face of the matter, to be no good reason why arms manufacturers should not exercise the right of every citizen to denounce measures he believes to be inimical to his interests and to those of employees and partners or shareholders.

Even so, the allegation that armament firms exerted a baleful influence on national policies before and after the First World War cannot be summarily dismissed; nor can the allegation that they formed international alliances in defiance of their obligations towards their own governments. It is also pertinent to ask how far firms organized on a commercial basis have succeeded not merely in earning profits for themselves or their shareholders but in forwarding national interests. To what extent have governments been able to count on them as suppliers of weapons needed in war or to meet national emergencies or serve diplomatic ends?

TWO

Beginnings

Some twelve centuries ago, Charlemagne forbade the export of armour from the Frankish dominions. These, if we include Slav lands whose rulers paid him tribute, covered practically the whole of modern Europe as far east as Bohemia and Croatia, except the Scandinavian countries, Great Britain, Brittany, southern Italy, Portugal and most of Spain. But this does not mean that there existed, in that remote era, an international arms trade whose activities were deemed inimical to the interests of the state. In the first place, the concepts of nationality and the state were foreign to early medieval thought. Secondly, the edict was pronounced in a context which suggests that it was aimed not so much at armourers as at vassals who might try to evade their military obligations on the pretext that they or their retainers were inadequately equipped.

The emphasis on armour is significant. At the battle of Adrianople in A.D. 378 about 45,000 Roman troops, led by the Emperor Valens, had suffered a calamitous defeat at the hands of roughly the same number of Goths because their infantry was unable to escape encirclement by armoured horsemen wielding swords and lances. Nearly two hundred years later a Frankish horde had been routed at Capua by Byzantine mounted archers. The conclusion drawn from these events by medieval strategists was that no commander of troops could afford to be without a substantial force of armoured cavalry.

To provide himself with such a force, virtually every medieval ruler in Western Europe instituted, at one stage or another, some form of military service which could be broadly described as feudal. At any rate in theory, every such ruler already possessed the right to 'call out the host', or in other words to summon all able-bodied freemen to his service in time of war and to require each man to bring with him rations, clothing and an appropriate weapon. The essence of feudalism was that some of those liable to military service agreed, in return for benefits conferred on them or their families, to come to the host not only armed but also mounted and armoured. Since only men of means could undertake this obligation and had time to master the art of managing a horse while encumbered with weapons, shield and body armour, service in the cavalry came to be associated with superior social status and, where the obligation was hereditary, with noble or gentle birth.

Experts have traced the origin of feudal practices both to the granting of leases by some Roman landlords in return for personal services instead of rent, and to the habit of Celtic and Germanic chieftains of surrounding themselves with privileged retainers who received food, shelter and protection as the price of their support in peace and war. To understand the influence of feudalism on the supply and procurement of arms, it will be enough to consider the adoption and development of these practices in the part of Western Europe ruled by the leaders of the West Franks and later by the kings of France.

In general, age-old custom among the Franks decreed that every household of free persons should contribute to the host, when called upon, one fighting man with his arms and equipment. Originally there seems to have been no limit to the length of time for which a man could be required to serve, but later the maximum period of obligatory service was often limited to as little as forty days. This method of recruitment provided a force of lightly-armed infantry suitable for only brief campaigns, organized in each district under the command of a count.

About the beginning of the eighth century, the introduction of the stirrup increased the effectiveness of cavalry by making it possible for a horseman to use a lance as a lever to unseat his opponent, or to rise in the saddle to deliver a crushing blow with a sword. Almost simultaneously, the Frankish dominions were threatened by the appearance of Moorish invaders north of the Pyrenees.

Charles Martel or Charles the Hammer, in effect though not in theory ruler of the Frankish kingdoms of Neustria and Austrasia, decided to meet the threat from the Moors by strengthening his cavalry. But the Frankish Crown had no revenue apart from the proceeds of the royal estates. Charles attracted recruits for his *corps d'élite* of armoured horsemen not by offering payment in coined money but by giving suitable candidates grants of land, called benefices, and by forcing reluctant bishops and abbots to use church lands for the same purpose. These grants enabled the recipients both to support themselves and their families and to equip themselves for war.

This was a feudal system inasmuch as holders of benefices were required to take an oath of absolute fidelity and were called vassals. But it lacked an essential element of what was afterwards understood by feudalism. Benefices were granted by Charles and his immediate successors solely to enable beneficiaries to perform specific tasks. They could not be passed from father to son, and they could be revoked if the beneficiary neglected his duties or became incapable of performing them. Church lands used as benefices remained, in theory, the property of the Church, although in practice control of them passed to the Crown.

In A.D. 732 Charles gained a decisive victory over the Moors near Poitiers. His cavalry were armed with throwing-axes and heavy swords. They were employed as mounted infantry, riding to the field of battle but fighting on foot. The Moors used mounted bowmen. They began by discharging arrows at ranges beyond the reach of throwing-axes. Had they persisted in these tactics, the Franks might have suffered much the same fate as their forefathers at Capua, although probably their armour would have saved them from such heavy losses. As it was, they were able to use their axes with devastating effect when the enemy closed the range, and in hand-to-hand combat their swords proved more than a match for scimitars.

Charles and his successors went on to develop the cavalry arm as a powerfully-equipped mobile striking force imbued with high standards of personal courage and devotion to duty. Throwing-axes were abandoned in favour of arrows. The lance became the principal weapon, but swords were still carried for close combat. Charles Martel's grandson, Charlemagne, besides banning the export of armour, decreed that every holder of a substantial tract of land should possess a shirt of mail and that every horseman should provide himself with shield, lance, sword, dagger, quiver, bow and arrows. He also took steps to improve the quality of the infantry levy by abrogating the rule that every free household must provide and equip one foot soldier. Holders of more than a stipulated area of land were required to serve in person. Holders of smaller properties were relieved of this obligation, but were ordered to club together to provide one man for every group of two or more holdings whose combined area reached the stipulated figure.

From the point of view of the supply and procurement of arms, the system reflected in these decrees had the advantage of putting the onus of providing the expensive equipment needed by heavy cavalry on the individual warrior. The supplier of arms and armour was not, like the modern arms manufacturer, a contractor angling for orders from governments, but a master-craftsman free to offer his goods in a wide market. On the whole, this method of procurement worked quite well as long as weapons continued to be hand-made and the wider implications of a feudal or quasi-feudal system to be compatible with prevailing social and economic conditions. The warrior, once he had accepted the obligation to serve, had every inducement to provide himself with the best weapons and armour he could get, not only because his life might depend upon them but also because they were status-symbols. The supplier had the satisfaction of exercising his craft in the knowledge that he was catering for a discriminating public.

The feudal system was, however, destined to reach its full flowering in Western Europe in circumstances which confronted the late Frankish and early French kings with formidable problems. Within thirty years of Charlemagne's death the empire into which he had welded the Frankish kingdoms fell apart. After a number of bloody disputes his grandsons agreed at Verdun in 843 to divide it into three parts, but each of them continued to think of himself as rightful King of the Franks, if not Emperor of the West. Weakened by dissension and assailed by Slavs, Saracens and Norsemen, they lost the power to revoke benefices once these had been granted. They were unable to prevent their vassals from forming private armies by themselves making grants of land in return for promises of support in peace and war. Every holder of a benefice, eager to affirm his social status, claimed the right to fight on horseback. Charles the Bald, by the terms of the Partition of Verdun King of Neustria with Aquitaine and the Spanish March, did not merely condone but positively encouraged this tendency by decreeing that every man who owned a horse or was in a position to do so should serve in the cavalry. Thereafter the cavalry tended to become not so much a band of dedicated warriors as an assemblage of proud and power-hungry aristocrats and their henchmen. By the end of the century benefices had become heritable fiefs, and nearly all important posts in the public service were held by men who claimed the right to bequeath their offices to their eldest sons or, if they were celibate, at any rate to nominate their successors.

Rulers who liked to think of themselves as the heirs of Charlemagne thus came to depend for much of their armed strength on forces controlled by noblemen on whose loyalty they had only an uncertain hold. The last of Charlemagne's male descendants in the direct line was killed in a hunting accident in 987. Hugh Capet, hereditary Duke of France and Count of Paris, was thereupon elected King of the West Franks and became in the eyes of posterity the first King of France. In theory he acquired the right to summon every able-bodied freeman to the host. In practice, he could seldom be sure that his orders would be obeyed outside the territories he held by right of inheritance. Fortunately for him and for the stability of the monarchy, these included Paris and extensive domains in the Ile de France and the Orléanais. He and his successors had none the less to meet repeated challenges from powerful nobles who repudiated or disregarded their feudal obligations.

That they survived not only these but also many other dangers says much for their resourcefulness, toughness and resilience. Even so, they could scarcely have done so had they depended solely on the feudal levy for their troops. The help which vassals who consented sooner or later to pay homage to the king were obliged to give was often so meagre as to be almost derisory. In the middle of the twelfth century the Count of Champagne contributed only ten knights to the host.[1] The Count of Flanders, with a thousand mounted men at his disposal, was not required to contribute more than twenty. It was chiefly by raising recruits in their hereditary domains, and by using money contributed by urban communities in lieu of personal service, that medieval French monarchs succeeded in forming armies capable of more than desultory warfare. In 1214 Philip II and his son Louis, facing a double threat from English and German invaders, managed to find seven thousand mounted troops for a campaign which culminated in a resounding victory for Philip at Bouvines. This was perhaps ten times as many as they could have mustered had they relied entirely on forces contributed by vassals. About five thousand of the seven thousand were not knights but sergeants or men-at-arms. These were volunteers drawn mostly from a class of small proprietors or yeomen who did not aspire to the privileges of knighthood and were content to do without some of its trappings. They were paid, but they did not serve primarily for pay or hire themselves in bands to the highest bidder.

In England there existed at the time of the Norman Conquest a system whereby the whole country was divided into a large number of more or less arbitrarily defined units, each responsible for providing, on demand, one foot soldier and his subsistence for sixty days. This method of recruitment was intended, like its Continental counterparts, to ensure that, although every able-bodied freeman was liable for military service, only an acceptable proportion of the adult males in a given community would be called up at one time. Thegns and eorls, as holders of land by feudal or quasi-feudal tenure, were obliged to serve whenever the host was called out. From the early part of the eleventh century, mercenary troops contributed a full-time professional element. Most of the population was rural, but towns and ports provided their quota of men for service on land or at sea.

Attempts by the Normans to establish a system more in keeping with the requirements of a permanent military occupation were not very successful. Settlers from Normandy who received large grants of land were required to build and maintain castles for the purpose of keeping the English in subjection and the Welsh at bay. In the long run these privileged immigrants and their descendants proved far more troublesome than the natives. Recalcitrant barons had repeatedly to be besieged in their strongholds. Massacres and expropriations had so drastically reduced the number of men available for military service that the host had always, or nearly always, to be supplemented on these occasions by expensive mercenaries. Furthermore, a serious disadvantage of the feudal system, in a country more than once threatened with invasion by sympathizers with rebellious vassals, was that it provided no permanent organization for defence. This weakness led to some curious *ad hoc* arrangements, such as the appointment of the Archbishop of Canterbury to concert anti-invasion measures in the south-eastern counties. Its ultimate consequence was that the Crown fell back on the expedient of entrusting the organization and control of its armed forces not to territorial magnates of doubtful reliability but to local officials, the shire-reeves or sheriffs. The holders of these posts were not vassals and their offices were not hereditary. Sheriffs were appointed by the king and could be dismissed by the king.

Armoured cavalry played a dominant role in European warfare for a thousand years or more. Even so, the cavalry arm proved by no means invincible. At the battle of Hastings Harold's mounted infantry, standing behind a rampart of shields on high ground six miles from the coast, came close to defeating the Conqueror's horsemen, admittedly not fighting in the most favourable conditions. In Asia Minor the Crusaders were so persistently harried by Saracen skirmishers that they had to use dismounted troops to support their cavalry. At Morgarten in 1315, and later at Sempach and Laupen, Swiss halberdiers and pikemen showed that well-trained infantry could stand up to cavalry and could even be used in a mobile role. Long before firearms became effective weapons of war, a cavalry charge could be stopped not only by halberdiers and pikemen but also by bowmen. At Crécy in 1346 English and Welsh archers, supported by spearmen and dismounted knights and men-at-arms and protected by a line of stakes stuck slantwise in the ground, broke up repeated attacks by French cavalry and played the chief part in routing an army with a numerical superiority of three to one. Using the same weapons and much the same tactics, the English were equally successful at Poitiers in 1356 and at Agincourt in 1415.

The resounding success achieved by the English in these battles has been widely attributed to their exclusive possession of the longbow and the addiction of English yeomen to archery as a sport or pastime. But Continental armies, though without the longbow, did have the crossbow, by no means an ineffective weapon. The crux of the problem was that good use could be made of the crossbow, the pike and the halberd only by specialist troops, and that these were usually volunteers or mercenaries whose services were not always readily available. Most European armies continued for some two centuries after Crécy to rely largely on the cavalry arm for their striking power, even though it had to be supported to a growing extent by other arms.

Since a medieval or early Renaissance knight would no more have thought of appearing on the battlefield without armour than a cricketer would think of going in to bat without pads and gloves, the result was that neither the success of the English and Welsh archers nor the adoption of such firearms as were available in the fourteenth and fifteenth centuries caused the armourer's art to fall into disfavour. On the contrary, new or improved weapons brought demands for even stronger protection for man and beast. Armourers responded to these demands by fashioning ingeniously jointed suits of plate armour which covered the knightly warrior from top to toe, but they had some difficulty in providing adequate protection for his steed. Good plate armour would withstand blows from pikes and the impact of projectiles from most portable firearms in use before the second quarter of the sixteenth century.

It was, however, very heavy. The wearer was apt to find himself at a disadvantage if he had to fight on foot because his horse was incapacitated; if thrown, he might have difficulty in rising to his feet. Moreover, an excess of ironmongery tended to impair the mobility of even the best-mounted cavalry. To carry a medieval knight and his half hundredweight or more of accoutrements, European horsebreeders had evolved a race of large, hairy-heeled chargers resembling nineteenth-century draught horses. These were considered a great improvement on the ponies or small horses ridden by barbarians and Roman knights alike in the Dark Ages. But speed was not their strongest point, and the growing weight of body armour towards the end of the age of chivalry threatened to make the cavalry even slower than it already was. Faster, lighter horses could give mounted troops a new lease of life as scouts and skirmishers, but only if they consented to rid themselves of some of their impedimenta and rely largely on speed and power of manoeuvre for their safety.

A further disadvantage of late medieval and early Renaissance armour was that it was very expensive. The high cost of fitting out a body of heavy cavalry was supportable as long as it was borne by warriors for whom the best equipment money could buy was both a life-assurance policy and a badge of rank. By the early part of the fifteenth century, however, the shortcomings of a method of recruitment which put the sovereign at the mercy of his vassals were so glaringly apparent that its abandonment in theory as well as practice could only be a matter of time. In 1439 King Charles VII of France struck a blow at what was left of the feudal system by forbidding the nobility to engage in private wars or to raise troops without his express approval. At the same time he announced the setting up of a regular army of permanently embodied troops, to be paid for from the proceeds of a special tax. These measures, stoutly resisted by the nobility but carried through despite their opposition, foreshadowed a state of affairs in which the whole burden of raising, equipping, training, supplying and paying troops would fall upon the state. The days were not far off when monarchs, unable to persuade their subjects to come to their aid by submitting to taxation, would have to sell or pawn their plate before they could go to war.

A step or two in that direction had already been taken in some countries. Mechanical engines of war which hurled stones, darts or other missiles at the enemy had existed since very early times. They began about 1326 to be supplemented, and were ultimately to be replaced, by firearms which used the explosive effect of black powder burning in a confined space to propel the missile. The earliest known drawing of a weapon of this kind depicts it as firing a feathered bolt, or quarrel, but afterwards cannon-balls of stone or iron were used. Since feudal warriors could not be expected to provide such weapons and they could not be readily improvised, their introduction pointed the way towards a shift of responsibility for the procurement of arms from the individual soldier to the state.

Ordnance was long believed to have been invented by Berthold Schwarz or Bertholdus Niger, a supposed Franciscan monk of Freiburg-im-Breisgau. However, some fifty years ago the English historian Sir Charles Oman showed that an alleged reference in the municipal records of Ghent for 1313 to the introduction of guns by a German monk did not occur in the original manuscript but had been interpolated in a copy made at a later date. No one knows who was the first gunmaker. The first guns of practical utility appear to have come from Flanders or Brabant, where Mons and Liège were early centres of production. They were made of wrought-iron bars wound round a wooden mandrel, hammered together and secured by wrought-iron hoops. This method of construction continued for some centuries to be used for large guns. Smaller guns could be made by the same method but were often cast in bronze, then commonly referred to as brass. Casting in iron needed higher temperatures than most early gunfounders could command.

Until the wheeled gun-carriage and the trunnion were introduced in the fifteenth century, guns were of little use for the support of troops in the field. There was, however, a keen demand for them as siege weapons. European sovereigns hastened to order guns from Flemish gunfounders, and some also imported men who could make them. The Plantagenet and early Tudor kings of England encouraged master-craftsmen from the Low Countries to settle in London or open branches there. Between the year of Crécy and the end of the fifteenth century at least a dozen gunfounders, among whom Peter of Bruges, William of Aldgate and John Cornwall were pioneers, transacted business in London or elsewhere in England.[2] One of them, William Woodward, sold no less than seventy-three guns to the Crown between 1382 and 1388.

However, even as late as the second decade of the sixteenth century domestic output did not suffice to meet all requirements. In 1512 Henry VIII bought sixteen large and twelve smaller guns from Hans Poppenruyter of Malines. He also ordered fourteen Flemish draught mares for each large gun.

In England, as in other countries, commercial gunfounders were not allowed to monopolize the secrets of their craft. Henry VIII followed the example of his predecessors by encouraging Flemish gunfounders to set up in business in England. But he also appointed two experts of foreign extraction, Peter Bawd or Baude and Peter van Collen or Peter of Cologne, to make experiments on his behalf. For two centuries and more, commercial gunfounders supplied practically all the guns for the armed forces. But the Crown, besides paying officials who gave technical advice, laid down standards and supervised tests, always retained at least the potential capacity to compete with private enterprise by ordering the construction of ordnance in royal arsenals.

Where shipbuilding was concerned, the Crown played a more active role. From the time when warships were first differentiated from merchant vessels, it became the custom for some to be built in royal dockyards and others at commercial shipyards. The same tendency to steer a middle course between monopoly by the state and the risk of monopoly by interlocking commercial enterprises has persisted right up to the present day in most arms-producing countries not wedded to Communist ideals. Arms factories throughout France were nationalized immediately after the French Revolution, but Napoleon soon reverted to the old system.

Early in the sixteenth century, arms manufacturers succeeded in developing a portable firearm which delivered a projectile capable of penetrating the stoutest body armour. This was the musket. Originally it was rather a cumbrous weapon which had to be supported on a rest. Even so, it could be carried by one man and fired from the shoulder. Muskets were used with some success at the battle of Pavia in 1525 and adopted soon afterwards by most European armies. A mixed force of musketeers and pikemen, the latter wielding inordinately long but well balanced polearms, was more than a match for the slow-moving cavalry of the day. The result was that the fully-armoured horseman with his cumbrous equipment disappeared from the European scene. Heavy cavalry, as well as light cavalry for reconnaissance and skirmishing, continued to be used, but it shed much of its ironmongery and became faster and more manoeuvrable. Body armour was gradually discarded, apart from helmets and breastplates retained at first for their practical utility and later as picturesque survivals.

About the same time, artillery became mobile enough to be used effectively on the battlefield. At the battle of Ravenna on 11 April 1512, Gaston de Foix and his ally the Duke of Ferrara, with a numerical superiority of five to two in guns, owed their victory over their Spanish opponents largely to the bombardment which preceded their frontal assault with infantry and cavalry.[3] Almost simultaneously, a growing awareness of the importance of sea power brought big demands for guns to be used at sea and for coast defence. These developments coincided with political, religious and dynastic reforms which widened the gap between England and Continental Europe and fostered among the English a lively interest in enterprises which promised handsome profits for the individual citizen and substantial benefits for the state. A prosperous gunfounding industry sprang up in the Weald of Sussex, where iron had been worked since Roman times and timber to make the charcoal used as fuel was plentiful. Besides making cast-iron cannon-balls and wrought-iron guns, Sussex ironfounders cast guns both in iron and in bronze, although cast-iron guns were still not very satisfactory. In Scotland, guns had been made at Edinburgh since the fifteenth century or earlier, and were afterwards to be made at Falkirk.

Gunpowder was made by servants of the Crown at the Tower of London as early as the first half of the fourteenth century, and later on the other side of the Thames at Rotherhithe. But it was also made by private manufacturers. The government factory at Waltham Abbey which closed in 1943 after serving Britain's armed forces in two world wars had been operating under private ownership for more than two centuries when the state acquired it in 1787. In 1590 George Evelyn, grandfather of the celebrated diarist, was granted a licence to install plant for the manufacture of gunpowder at two places in the south of England. The East India Company's interests, when its charter was renewed in 1693, included a powder-mill in Surrey. The Crown bought in 1759 a powder-mill established by private enterprise at Faversham, in Kent, but sold it in 1825 to a commercial firm, John Hall and Sons. As a result of mergers and takeovers, it passed through various hands and eventually became the property of Imperial Chemical Industries Limited. Many smaller powder-mills sprang up in England during the Napoleonic Wars but afterwards went out of business.

At the time of England's quarrel with Spain and for some years after it, the proximity of the Sussex foundries to South Coast ports gave rise to the fear that arms and ammunition might be 'sold out of the realm' and pass into the hands of hostile or potentially hostile powers. Local representatives of the Crown were warned to be on their guard and the export of ordnance was forbidden except under licence. In 1619 the Privy Council imposed further restrictions. Gunfounding elsewhere than in Sussex and Kent was prohibited. Guns were to be sold and tested only in London, and guns sent abroad were to be shipped exclusively from Tower Wharf. All guns were to be inscribed with the name of the founder, the weight of the piece and the year of manufacture. But some of the Sussex gunfounders were strongly suspected of evading these regulations, and a prominent citizen of Lewes was heavily fined for doing so.

During the seventeenth century greatly improved flintlock

muskets, effective at eighty to a hundred paces (though often used at longer ranges), superseded in most armies the wheel-lock or primitive flintlock muskets of the preceding century. At the same time King Gustavus Adolphus of Sweden opened up new prospects for gunners by making his artillery more mobile and improving its organization. One effect of these and other changes was that warfare grew steadily more expensive and wars more difficult to finance. The recently-formed Dutch Republic found an answer to this problem when, in 1609, the creation of the Bank of Amsterdam made it possible for the state to raise long-term loans on the security of the taxes and thus maintain armed forces which matched or surpassed those of larger and more populous countries whose finances were not so well organized. Nearly ninety years elapsed before the English applied the lesson by setting up the Bank of England. The French waited until their country was torn by revolution before setting up the Bank of France.

In the meantime the design and manufacture of ordnance made only slow progress. Cast-iron guns were cheaper than bronze guns. Also, more rounds could be fired in quick succession from an iron than from a bronze gun before the metal became dangerously soft. But the barrels of cast-iron guns were apt to crack, and in any case their production consumed such vast quantities of charcoal that only countries with almost unlimited supplies of timber could afford to go on making them in large numbers for an indefinite period if timber was also to be used for other purposes. The English had begun as early as the reign of Elizabeth I to replace timber by coal for the heating of houses and for some manufacturing processes, but this was possible only in places close to coal mines or to which coal could be carried by sea. Many attempts were made to use coal instead of charcoal for the smelting of iron ore, but little success was achieved before the eighteenth century. Even then the cast-bronze gun (still called the 'brass' gun in British military parlance) was not destined to be ousted from the apparatus of European armies for a long time to come. Progress in metallurgy might have been expected to bring radical changes in design, but it failed to do so. Gunfounders continued to make the weapons they were asked to make. In most cases these differed only in detail from those with which artillerists, qualified by seniority for responsible positions in ordnance departments, had been familiar in their youth.

Sweeping changes in the equipment of armies had, therefore, yet to come when Europe was convulsed by the French Revolution. Even so, the revolution and its aftermath of international conflict gave the world a foretaste of the methods by which nations would prepare themselves for wars of survival in future years. A national levy enabled the new rulers of France to raise armies on a scale never previously attempted. Men distinguished in various walks of life were called in to advise them how to increase output and find substitutes for scarce materials. Semi-skilled workmen, using simple hand tools but provided with moulds and jigs, turned out muskets in quantities hitherto unknown. Commercial and industrial enterprises deemed essential to the war effort were taken over by the state.

Among these was a foundry near the villages of Montcenis and Le Creusot, in a corner of Burgundy where iron ore was found in close proximity to deposits of coal mined since the early years of the sixteenth century. Shortly before the revolution its owners had employed James Watt to install steam engines and William Wilkinson to put into operation his system of re-melting cast iron with coke.

Napoleon returned the foundry to private ownership soon after he became First Consul. For some years it brought in a handsome revenue by producing bronze and cast-iron guns and a variety of projectiles. But its prosperity did not survive the restoration of peace. Between 1815 and 1835 the business passed through the hands of at least four different proprietors, two of whom went bankrupt.

In 1836 Eugène Schneider, a Frenchman whose family came from the Saar, bought the derelict ironworks at Le Creusot and formed, in association with his brother Adolphe, a company called Schneider Frères et Compagnie.[4] The name was changed to Schneider et Compagnie when Adolphe died in 1845 and Eugène became sole proprietor. The Schneider brothers had an uncle who was an officer in the French Army and afterwards became Minister of War. But there is no reason to suppose that they foresaw, when they made the purchase, that the company they founded would become one of the world's largest arms-manufacturing concerns. Their immediate aim, perhaps their sole aim, was to profit by the railway and steamship boom. Besides installing plant for the manufacture of locomotives and rolling stock at Le Creusot, they opened a marine department not far away at Chalon-sur-Saône. They completed their first locomotive in 1838, their first steamboat in 1839.

On the far side of the Atlantic, the nucleus of a gunfounding industry existed in North America in colonial times, and there was a keen demand for small arms from landowners, sportsmen and prospectors. German and Swiss settlers in Pennsylvania played leading parts in developing fowling-pieces with which, it was claimed, a skilled marksman could hit a turkey at two hundred paces or its head at sixty to a hundred. Local committees set up in 1775 to discuss the procurement of arms for the militia found that there were some two hundred gunsmiths in the thirteen colonies, most of them in Maryland, Massachusetts, Pennsylvania and Rhode Island. The small arms in service on the outbreak of the War of Independence were chiefly smooth-bore muskets with calibres in the region of half an inch. The Marquis de Lafayette arranged for the shipment from France of large numbers of Charleville muskets of a pattern adopted in 1763. These took their name from one of three leading small-arms factories established early in the eighteenth century, the others being at Maubeuge and Saint Etienne. Production of an American-made version of the Charleville Model 1763 began at Springfield Armory in 1795, and an improved model was introduced in 1797.

Rifled muskets, commonly called rifles, were adopted in most armies about the beginning of the nineteenth century, but at first were issued only to specialist units. During the Napoleonic Wars the British Rifle Brigade used a very good rifled musket designed by Ezekiel Baker of Whitechapel, but the Baker rifle was afterwards replaced by the less satisfactory Brunswick Jäger rifle. This was a large-calibre version of a rifled musket used by the Hanoverian Army. Besides being difficult to load, it was notorious for its heavy recoil.

Difficulty in loading was one of two major weaknesses of most rifled muskets. The other was that the rifling was apt to become clogged after a few rounds by the waste-products of combustion. Despite the invention of guncotton and nitro-glycerine in 1838 and 1845 respectively, little could be done about the second weakness until smokeless powders made from these materials were introduced many years later. Ingenious attempts were, however, made to overcome the difficulty of ramming into the barrel of a rifled musket a projectile wide enough to fit it snugly. These culminated in the 1840s in the introduction by the French of a hollow conical bullet—the Minié bullet—which was slim enough to be pushed easily into the barrel but expanded when the charge was fired. The British decided in 1851 to adopt as a standard infantry weapon a rifled musket designed to accept a large-calibre version of the Minié bullet, but only about three and a half thousand British-made Minié muskets were manufactured by the end of 1853.

Another way of countering objections to the muzzle-loading rifled musket was to dispense with muzzle-loading. Many attempts were made in the first half of the nineteenth century to design breech-loading sporting guns and military rifles which would combine ease of loading with a gastight seal and a reliable means of igniting the propellant. John Pauly, a Swiss working in Paris, experimented with ignition by a primer which detonated when a needle pierced a priming cap. In 1816 he lodged a specification which foreshadowed the use of a cartridge made wholly or partly of ductile metal to seal the breech of a breech-loading handgun. Twenty years later E. Lefaucheux introduced a hinged breech-loading gun which accepted a paper cartridge with a metal base. A pin projecting from the side of the case detonated the primer when it was struck by a falling hammer, but despite its metal base the cartridge did not provide a gastight seal. By adopting an improved firepin cartridge proposed by an associate, Lefaucheux afterwards succeeded in producing a double-barrelled breech-loading gun which was the forerunner of all modern sporting guns.

In the meantime Nicholas Dreyse, a German who had worked for Pauly in Paris, introduced a breech-loading 'needle-gun' which employed a turn-bolt mechanism. A paper cartridge held the powder, and the primer was attached to the back of the bullet. When the trigger was pulled, a long, slender needle passed through the cartridge and struck the primer. Serious disadvantages of this arrangement were that the cartridge did not expand to seal the breech and that the needle was subject to corrosion and sometimes broke. Nevertheless Dreyse's breech-loader was adopted by the Prussian military authorities and afterwards widely copied.

The 'needle-gun' made its first appearance in 1838. Ten years later Christian Sharps, a citizen of the United States, improved on Dreyse's device by using a knife-edge to cut the cartridge when the bolt was closed, so that pulling the trigger caused the primer to flash straight into the powder. The Sharps rifle of 1848 is said to have been the first in which the breech was effectively sealed, but the paper cartridge cannot have been entirely satisfactory since a metal cartridge was used in later versions.

Another trend-setting American weapon of the period was the Jennings repeating rifle patented by Lewis Jennings in 1849. Hollow conical bullets were used, and the barrel had to be tilted upwards between shots to allow the next round to fall into place. The Jennings rifle, manufactured by Robbins and Lawrence of Windsor, Vermont, was one of the ancestors of the Smith and Wesson, Volcanic, and Winchester rifles (but not Smith and Wesson pistols) which afterwards achieved world-wide fame. These, however, incorporated a double-toggle lock-joint which was invented not by Jennings but by Horace Smith and Daniel B. Wesson, whose patent was taken out in 1854.

So the first half of the nineteenth century was an era of substantial progress in the design of small arms. It was also an era of substantial progress in methods of production. At the very beginning of the century Eli Whitney, taking as his point of departure a description by Thomas Jefferson of methods employed in revolutionary France, used not only moulds and jigs but also water-powered machinery to manufacture ten thousand muskets for the United States government. About eighty per cent of the plant used later by Samuel Colt to turn out the carbines and pistols which made his name a household word was mechanized. By the middle of the century the components of his and some other American weapons were produced to standards which made them interchangeable. Robbins and Lawrence showed at the Great Exhibition in London in 1851 not only complete rifles but also interchangeable parts.

Nothing like the same progress was made in the design and production of ordnance. In 1850 most field artillery in all armies was equipped with smooth-bore cast-bronze guns whose characteristics had changed little since the middle of the preceding century. The theoretical advantages of a rifled barrel were well known to artillerists, but bronze guns were not considered suitable for rifling. Experiments with rifled barrels were confined to large guns intended for siege warfare or the defence of fixed positions. In Britain the Master-General of the Ordnance and his staff were responsible for the design of all guns for the army and the navy. Independent experts were sometimes consulted about specific problems. but they were not encouraged to concern themselves with the wider aspects of research, development and design. Commercial gunfounders employed as government contractors worked from drawings supplied by the Ordnance Department.

Furthermore, in the middle of the nineteenth century not one of the men who were to achieve international fame as arms manufacturers during the next fifty years or so had yet received an order for as much as a single gun. The principal suppliers of cast-iron ordnance to the British government were the Low Moor Company of South Yorkshire and Walkers of Gospel Oak, Staffordshire. Eugène Schneider was still making locomotives, steamboats and marine engines. Thorsten Nordenfelt and Hiram Maxim had yet to invent the weapons that would make them famous. Joseph Whitworth was turning out machine-tools, gauges and industrial machinery which set new standards of precision. William Armstrong, bred to the law but a born engineer, had set up as a manufacturer of hydraulic cranes and general engineering products. Edward Vickers was senior partner in a Sheffield firm which was making handsome profits by exporting steel bars and sheets to the United States. Alfried Krupp, who had taken to spelling his first name Alfred after a visit to England in 1839, was a successful steel magnate with a special interest in railway material. He had made and tested at least one gun, but had yet to sell one. Nor did he receive any orders for guns when he showed, at the Great Exhibition in London in 1851, a small all-steel cannon flanked by a monstrous steel ingot.

THREE

The New Look

Soon after the middle of the nineteenth century, a quarrel between Roman Catholic and Greek Orthodox monks about access to the Church of the Holy Nativity at Bethlehem led to hostilities between Turkey and Russia. Since the Tsar Nicholas I had made no secret in recent years of his designs on the supposedly moribund Ottoman Empire, his espousal of the cause of his co-religionists in Asia Minor aroused deep misgivings in Western Europe. Britain and France, not content with giving the Turks diplomatic support, declared war on Russia. At a Cabinet meeting one evening in the summer of 1854, when the weather was so hot and the hour so late that some ministers are said to have been too sleepy to know what was going on, the British government committed itself to an Anglo-French expedition to the Crimea, with the Russian naval base at Sebastopol as its objective.

The Crimean War was the first fought between major European powers since the coming of the railways, the steamship and the electric telegraph. The technical developments of the past forty years or so had, however, made little impact on European armies. In 1854 British and French troops wore uniforms more suitable for the parade-ground than for an overseas campaign. Their artillery consisted almost entirely of old-fashioned smooth-bore muzzle-loaders, supplemented by a few siege guns with experimental rifling. The French infantry was armed with rifled muskets which used the Minié bullet. The British still relied largely on smooth-bore muskets only marginally different from those which had served them well since the last decade of the seventeenth century, but enough of the new Minié muskets were available by the end of the year for some 34,000 to be distributed to units in the field. The British Crimean Expeditionary Force had no ambulances, and only the most meagre provision for the care of the sick and wounded was made above regimental level until Florence Nightingale prodded the War Office into action. The French did have ambulances, but they were handicapped by outmoded tactical doctrines and the disinclination of their leaders to depart from principles laid down when Napoleon's armies were sweeping irresistibly through Europe.

As for the Russians, they had already shown themselves incapable of overcoming even the allegedly incompetent Turks. They relied chiefly on smooth-bore muzzle-loading weapons, but they had some rifled muskets designed to accept large conical bullets similar to those used by the British, and their 12-pounder field guns outranged the 9-pounders used throughout the campaign by the British Field Artillery. They were hampered at times by an extreme reluctance to allow guns to fall into the enemy's hands, but were helped by having as their strategic aim the retention of an objective which proved fairly easy to defend against opponents slow to take advantage of their opportunities.

They were also helped by difficulties sometimes experienced by the British and the French in concerting their movements, and perhaps too in other ways. The Crimean campaign was the first in which a British Expeditionary Force was accompanied in the field by an accredited correspondent of a national newspaper. W. H. Russell of *The Times* sent home a stream of despatches containing a good deal of criticism, implicit or explicit, of the Allied military authorities. These were published with additional material contributed by the staff in London. Russell and his editor performed a useful service by drawing attention to the shortcomings of the War Office, but their disclosures did nothing to ease the task of the British Commander-in-Chief in the field, Lord Raglan. On the contrary, Russell sometimes embarrassed Raglan by transmitting accounts of his dispositions and intentions which, although not always accurate, included information that might be useful to the enemy. 'We have no need of spies,' a Russian general was reported as saying. 'We have *The Times*.'

The Allies had command of the sea, but lacked the well-protected vessels of shallow draught they would have needed to bombard Sebastopol at close quarters in face of return fire from shore batteries. They decided that their troops should land on the west coast of the Crimea and advance overland to the objective.

Bombardment vessels of unorthodox configuration, said to have been designed by the Emperor Napoleon III in consultation with the naval architect Stanislas Dupuy de Lôme, were ordered by the French naval authorities for eventual use at Sebastopol, and Eugène Schneider made his début as a manufacturer of armaments by supplying engines and armour for them. He also supplied engines for gunboats, frigates and larger warships. An aspect of these transactions which made a particularly favourable impression on the French government was that seventeen engines were delivered within seven months of the placing of the first order.

The British government also went outside the circle of its usual suppliers of armaments by inviting William Armstrong, on the strength of his reputation as an engineer with a special interest in hydraulics, to design underwater mines suitable for blowing up vessels sunk by the Russians as blockships. The mines he made and tested were never used for their intended purpose, but his first contact with the military authorities paved the way for a subsequent approach which was to have momentous consequences.

British and French troops, accompanied by a small Turkish contingent, landed without opposition in the Crimea on the appointed day. On 20 September, after forcing the crossings of the Alma, they made a leisurely advance to the extreme south-west corner of the Crimea, established bases there, and prepared to capture Sebastopol from the south while their warships watched the entrance to the harbour. But they omitted to block roads and tracks linking the port with the interior. The Russians were able not only to strengthen the defences of Sebastopol and reinforce its garrison, but also to build up a substantial force outside the town. By the end of October their forces in the Crimea outnumbered those of the Allies by nearly two to one in men and not far short of two to one in guns.

On 25 October the Russians made an unsuccessful attempt to cut communications between the British base at Balaclava and the Allied front outside Sebastopol. The 'thin red line' of the 93rd Foot withstood assault by four squadrons of cavalry; an uphill charge by the Heavy Brigade of the British Cavalry Division threw the main body of the Russian cavalry into confusion; and the Light Brigade suffered heavy losses while trying to comply with an ambiguous order carried by an unreliable emissary. The

Russians succeeded only in occupying some ground of no great importance.

On the day after Balaclava, six Russian infantry battalions supported by four field guns broke through the outpost positions of the British 2nd Division near the centre of the Allied line. But they were halted and dispersed by the division's bronze 9-pounders before they could reach the main position on a crest called the Home Ridge. This action was afterwards called Little Inkerman.

The real battle of Inkerman was a far more ambitious affair. It was fought on Sunday, 5 November. The Russians planned to march about 15,000 troops out of Sebastopol under cover of darkness and unite them on the left bank of the Tchernaya with approximately the same number of troops of their field army which were to cross the river by a bridge close to the ruins of the ancient settlement from which the battle took its name. Under cover of demonstrations and diversionary attacks on the Allied right, the combined force was then to drive straight through the centre of the Allied line with the object of ending the threat to Sebastopol by eliminating the British component of the Allied Expeditionary Force.

Rumours of an impending offensive reached the British some days before the attack was due, but patrols sent to the left bank of the Tchernaya and to positions commanding a view of the right bank saw no sign of any major change in the enemy's dispositions. On the eve of the offensive the 2nd Division still held the Home Ridge. Picquets were posted up to half a mile forward of the main position, and a road leading from the bridge across the Tchernaya was blocked at a point about 300 yards in front of the ridge. Routine precautions included the posting of picquets within earshot of the route which troops emerging from Sebastopol would follow.

Heavy rain began to fall during the afternoon of Saturday, 4 November. It eased up after midnight, but at dawn on Sunday the Home Ridge and the wooded gulleys in front of it were swathed in mist and drizzle. The jingle of harness and the rumbling of wheels were heard in the early hours of the morning, but these sounds were thought to come from market carts carrying produce to Sebastopol from the interior.

The alarm was given about 5.45 a.m., when an officer making

a routine visit to his forward troops heard shots exchanged between Russians coming from Sebastopol and picquets posted by the formation on the 2nd Division's left.

By boldly sending troops forward to meet the enemy in the gulleys below the Home Ridge, the acting commander of the 2nd Division succeeded in breaking up the cohesion of the force from Sebastopol and delaying its arrival at the rendezvous, but he could not prevent the enemy from bringing up more than a hundred guns, mostly from the far side of the Tchernaya.

As soon as visibility improved, these guns opened a damaging fire on the British lines. The 2nd Division's field guns replied, but the Russian artillery was close to the effective limit of their range.

Lord Raglan reached the forward area about 7 a.m. Seeing that the 2nd Division was outgunned, he gave orders that two 18-pounders with more than twice the range of the 9-pounder should be brought from the artillery park outside Sebastopol. But his message miscarried. When it did reach the right destination, a long time elapsed before the guns, which weighed with their accoutrements the best part of three tons each, could be brought forward and manhandled into position by volunteers.

The 18-pounders reached the neighbourhood of the Home Ridge about 9.30 a.m. They were joined soon afterwards by guns sent forward by the French. About half an hour later the enemy's artillery fire began to slacken. There was heavy fighting throughout the greater part of the day at the roadblock and elsewhere, but long before it was over the Russians tacitly admitted that the tide had turned against them by starting to pull back the field batteries they had moved forward with such high hopes at the outset of the battle. In the afternoon, after losing about three times as many killed and wounded as the British and the French, they withdrew from the battlefield without ever reaching the Home Ridge, although their infantry had made repeated attempts to storm it.

The battle of Inkerman was thus a tactical defeat for the Russians. Even so, their offensive was not altogether unsuccessful. The losses suffered by the British and the French on 5 November, even though they were much lighter than the enemy's, ended any chance the Allies might have had of capturing Sebastopol before the winter.

In the light of all that is now known about the battle, it seems clear that the part played by the 18-pounders was somewhat exaggerated in contemporary newspaper reports and that less than justice was done in British accounts of the fighting to contributions made by the French. Inkerman was a soldier's battle, fought in conditions which, once it was well launched, made effective control of formations by commanders and their staffs extremely difficult and which put a premium on the courage, initiative and enterprise of combatant troops from regimental commander to private soldier. The fact remains that the arrival of the 18-pounders, although not decisive, did mark a turning-point in the battle. In effect, the Russians conceded victory to the Allies when they saw that their guns in forward positions were in danger of being overrun by the enemy's infantry and started to withdraw them. But they must have known earlier that their attempt to make the British front untenable by exploiting the superior range of their field artillery had failed.

At the time, accounts of the battle were eagerly read in London. Among those who studied them with special interest were William Armstrong and his friend James Meadows Rendel, with whom he was staying at the time.

William George Armstrong was 43 years old when the battle of Inkerman was fought. Son of a prosperous corn-merchant of Newcastle-on-Tyne and grandson of a Cumberland shoemaker, he had been educated privately and at a grammar school at Bishop Auckland. His father had articled him when he left school to a local solicitor named Armorer Donkin, but he had spent three or four years studying the law in London before returning to Newcastle to become a partner in Donkin's firm. Always keenly interested in engineering, he had made friends while he was still a schoolboy with William Ramshaw, proprietor of a local engineering works, and in 1836 had married Ramshaw's daughter. As a practising solicitor he had found time not only to take a leading part in forming the company which gave Newcastle its first piped water, but also to invent a hydro-electric generator which attracted such favourable attention in scientific circles that he was elected a Fellow of the Royal Society. At the age of forty he had abandoned the law to found, with financial backing from Donkin and other friends, an engineering firm called the Newcastle Cranage Company. The firm specialized in hydraulic cranes but was more than willing to accept any orders that might come its way for other engineering products.

A portrait of Armstrong at the age of twenty depicts him as a strikingly handsome, slightly dandified youth with prominent but well-cut features and the air of romantic melancholy which the poets and novelists of the Napoleonic era had made fashionable. He was to retain his good looks almost to the day of his death at the age of ninety. Childless, though devoted to his friend Rendel's children, he had become in early middle age a compulsive worker who sometimes followed a day at the factory by spending the night there. When he did leave Newcastle it was usually to stay with Rendel in London and attend meetings of the Royal Society.

James Meadows Rendel was a distinguished engineer whose family was destined to make a great deal of money from the arms trade. According to his descendants he came of Devonshire farming stock. His father, however, was not only a farmer but also a surveyor. Himself trained as a surveyor, he seems to have learnt the rudiments of engineering from an uncle who owned a mill. By the time of the Crimean War he had made himself one of the world's leading authorities on the design and construction of docks and harbours. He had also become a rich man, able to give his sons the best education money could buy. Like Armstrong, he was a Fellow of the Royal Society. It was largely on his advice that Armstrong had given up the law to become a manufacturer of cranes, although Rendel had contributed no capital to the venture but contented himself with putting business in his friend's way.

On reading about the use made by the British of their 18-pounders at Inkerman, Armstrong and Rendel were struck by the length of time and the vastness of the effort needed to move two guns of moderate calibre from the artillery park to the vicinity of the Home Ridge, less than two miles away. Rendel felt that, if the government's ordnance experts were incapable of designing a light and handy gun with a better performance than the Russian 12-pounder, then the time had come for a civilian member of the engineering profession to show them how the problem should be tackled. 'And you,' he said to Armstrong, 'are the man to do it.'[1]

Up to that time Armstrong, so far as is known, had never designed or made a weapon of war apart from the underwater mines he had been asked to make at the outset of the campaign. Nevertheless he responded to his friend's suggestion by roughing out on a sheet of blotting-paper a design for a compact field gun based on the principle of the small-arms rifle.

He and Rendel then decided to turn the matter over in their minds before committing themselves to any further action. It was not until December, when the campaign in the Crimea was at a standstill, that they wrote to a senior official of the War Office, 'suggesting the expediency of enlarging the ordinary rifle to the standard of a field gun and using elongated projectiles of lead instead of balls of cast iron'.[2]

There was nothing very new in these proposals. Ordnance experts had long known that a rifled cannon firing an elongated projectile would have theoretical advantages over a smooth-bore cannon firing a spherical one. Apart from suggesting that a rifled musket might, in some unspecified way, be scaled up to the dimensions of a field gun, Rendel and Armstrong did not say how the practical difficulties involved in the making of such a weapon could be overcome. But a communication from two Fellows of the Royal Society could not be lightly brushed aside. Their letter was shown to the Duke of Newcastle, Secretary of State for War and the Colonies in a government soon to be driven from office by popular dissatisfaction with its conduct of the war. Armstrong saw the Duke, and was invited to carry out experiments on the lines proposed and to make up to half-a-dozen experimental guns.

He soon found that making even one gun that would meet his requirements was a formidable task. His head draughtsman, a Quaker named Richard Hoskins, declined to work on the project, saying: 'Thou knowest, Mr Armstrong, that I cannot go against my conscience.'[3] He had to try many different materials and methods of construction before he was satisfied that he could make a rifled barrel long enough to impart the required muzzle velocity to an elongated projectile and strong enough to withstand the forces imposed on it. He rejected cast iron and cast bronze as either lacking in tensile strength or too heavy. Cast steel offered the highest tensile strength in proportion to its weight of any of the materials he tested; but he decided against a barrel made from a solid mass of cast steel, on the ground that it was not yet possible to produce flawless masses of cast steel of the required size. The method of construction he adopted after many trials was to shrink on to a rifled inner barrel of steel or wrought iron a sleeve formed from a wrought-iron tube wound round a mandrel and welded at the edges. If necessary a second sleeve could be superimposed on the first, and the process could be continued until the thickness needed to give the required strength was reached.

Armstrong adopted the breech-loading principle for his experimental guns, and for most of his later guns except some naval guns and siege guns of large calibre. To close the bore he used a vent-piece, or block, which fitted into a slot and was held in place by a hollow screw through which the gun was loaded. Later the screw was replaced by a wedge.

This was the weakest feature of Armstrong's early guns. Even if the block did not blow out, as it sometimes did, it tended to give a less than gastight seal. What artillerists call poor obturation was, indeed, a shortcoming of most breech-loading guns produced during the next twenty years or so.

He also had some difficulty in producing a suitable projectile. Much more was involved than the mere substitution of 'elongated projectiles of lead' for cast-iron cannon balls. Shells filled with highly explosive materials such as trinitrotoluol and its fore-runners had yet to come, but even in the middle of the nineteenth century gunners expected to be able to use, as an alternative to solid shot, a hollow projectile filled with shrapnel or with com-bustible material. Eventually, and in the light of many hazardous experiments, Armstrong adopted a cast-iron projectile with a coating of lead just thick enough for the grooves of a rifled barrel to work upon it. He also devised an adjustable fuse which could be set to explode a shell either on or before impact. A shell fitted with this fuse was, he claimed, so safe up to the moment when the gun was fired that it could be thrown from the roof of a house without exploding.[4]

His first gun, a 3-pounder, was ready for delivery in the summer of 1855. It passed its tests, but was pronounced too small. The authorities returned it to Armstrong with the suggestion that he should rebore it for use as a 5-pounder. In the following year, when the Crimean War was over, he produced a 5-pounder which gave good results at more than twice the maximum range of the smooth-bore 9-pounder which was still the standard weapon of the field and horse artillery. At the end of another two years he submitted for trial an 18-pounder which proved to have a better performance at a range of two miles than the smooth-bore 18-pounder at half the distance.

Satisfied that Armstrong had made out a good case for rifled artillery, the government appointed in 1858 a Committee on Rifled Cannon to decide not whether rifling was desirable but what kind of rifled gun should be adopted.

By that time Armstrong had a serious rival in Joseph Whitworth, the brilliant manufacturer of machine-tools and industrial machinery from whom he had bought one of the first machines for his cranage works. Well known for his advocacy of standardization and precision, Whitworth had done valuable work for the authorities in connection with the design and construction of small arms. He had also produced rifled barrels for field guns from blocks of metal sent to him for the purpose. He had, however, not yet built a gun of his own, complete in all its parts, although he was ready by 1858 with a design for one.

In that year and the next, the authorities held repeated trials at which rifled guns submitted by six different designers or manufacturers were matched against smooth-bore guns. Whitworth was represented by one of the guns whose barrels he had rifled for the Ordnance Department.

The outcome was a triumph for Armstrong. The Committee on Rifled Cannon reported in the light of the trials not only that his entry was 'vastly superior' in range and accuracy to all existing field ordnance, but also that it had been offered 'no practical evidence' that any alternative system of rifling was as good as his. Lord Panmure, the new Secretary of State for War, added that Armstrong's gun was seven times as accurate at 3,000 yards as 'the ordinary smooth-bore gun' at 1,000 yards, and at equal distances was fifty-seven times as accurate. The Duke of Cambridge, holder of the archaic post of Commander-in-Chief of the British Army, said Armstrong's gun could 'do everything but speak'.[5]

Thereupon the government decided to place large orders for guns built to Armstrong's designs. To provide the necessary manufacturing capacity, a company called the Elswick Ordnance Company was established as an offshoot of Armstrong's engineering company at Elswick, in what were then the outskirts of Newcastle. In return for financial backing and a promise of steady employment as long as the arrangement held good, the new company pledged itself to manufacture ordnance solely on government account while its contracts with the government remained in force. George Rendel, one of three sons of James

Meadows Rendel who were associated at one time or another with Armstrong's projects, was appointed General Manager of the ordnance factory and became a partner in the undertaking. The other partners were George Cruddas and Richard Lambert, both Newcastle men and partners in the engineering firm since its inception.

Armstrong himself refused a partnership in the ordnance company, but retained his interest in the engineering firm. He presented the patents covering his guns, projectiles and fuses to the nation, and was rewarded with a knighthood and a well-paid government post. As Engineer of Rifled Ordnance, responsible for supervising production at Elswick and later also at the royal foundry at Woolwich, he received what was then the very large salary of £2,000 a year, back-dated to 1856. In addition he was reimbursed for his expenditure on research and development since 1855.

About three thousand guns, including some large muzzle-loaders, were built to Armstrong's designs between 1858 and 1863.[6] They were not exposed to the test of a European war, but some were used in small colonial wars and in the China war of 1856 to 1860.

Armstrong's renunciation of the royalties his patents might have earned him made him a popular hero, but he was not allowed to enjoy his triumph in peace. Joseph Whitworth was not a man who readily accepted defeat, and he had powerful supporters in political circles and among industrialists who saw no reason why they should not have been asked to build guns to Armstrong's designs. The government's decision to act as fairy godmother to the new-born Elswick Ordnance Company, instead of parcelling out orders among its usual suppliers or concentrating production at Woolwich, was hotly criticized. Whitworth claimed, perhaps not without reason, that he was capable of designing and manufacturing guns just as good as Armstrong's. He alleged—although here he seems to have been mistaken—that he had been excluded from the trials supposed to have proved the superiority of Armstrong's entry. Pointing out that the Committee on Rifled Cannon had visited Armstrong's factory but not his, he complained that the authorities had reached a decision unfavourable to him without giving due consideration to his proposals.

These were not the only reproaches levelled at the government.

The authorities were charged with putting Armstrong in a position to pass judgement on the work of rivals. Some critics accused them of squandering public money on Elswick-made guns which cost, they said, £2,000 apiece. Others alleged that Elswick-made guns sent to China had proved so unreliable that they had to be withdrawn from the firing line in haste.

All such charges were investigated at considerable length by government departments or quasi-judicial bodies. After the controversy had dragged on for five years or so, the government of the day set up a committee of enquiry with power to interrogate witnesses and bring all allegations into the light of day. When its composition was announced, Whitworth pointed out that it consisted entirely of serving officers of the armed forces. His request that he should be allowed to nominate an additional civilian member was granted, and Armstrong was invited to do the same. Whitworth chose an eminent designer and constructor of marine engines. Armstrong's choice fell on Stuart Rendel, a brother of George Rendel and a rising barrister with political ambitions. Unlike Armstrong, who was what would afterwards have been called a staunch Conservative, Stuart Rendel supported the radical wing of the Liberal party. An able advocate and a shrewd investor who mixed on terms of easy friendship with men prominent in public affairs, he is generally supposed to have been hampered in his political career by his association with the arms trade. But in due course he was rewarded for his services to the party with a peerage bestowed by Gladstone.

Most of the charges brought against the government were refuted without difficulty. Men whose integrity and knowledge of the facts were beyond dispute bore witness that the Committee on Rifled Cannon had reached its conclusions in the light of exhaustive trials at which Whitworth's entry competed on level terms with other entries. The Committee had visited Armstrong's factory and left Whitworth's unvisited not because its members were prejudiced against Whitworth but because, at the relevant time, he had yet to make a gun and inspection of his plant would therefore have thrown no light on the production problems they were investigating when they inspected Armstrong's. Captain Andrew Noble, who had acted as secretary of the Committee on Rifled Cannon and other ordnance committees, testified that he had invited Whitworth to every one of the trials and clearly remembered his attending some of them. True, he was the same Captain Andrew Noble who, by the time the Committee of Enquiry was appointed, had quitted the public service to work for the Elswick Ordnance Company; but his veracity and the accuracy of his recollection were not seriously challenged. The allegation that Armstrong had been put in a position to pass judgement on the work of rivals was shown to be founded on ignorance of the system in force. Deciding which of the designs submitted to the authorities should be accepted and which rejected was the business not of the Engineer of Rifled Ordnance but of the Ordnance Select Committee. Armstrong was not a member of that committee, and such questions were never referred to him. Indeed, in the light of experience in the Crimean War the organization of the department had been modified for the express purpose of ensuring that no one who submitted a design should ever be called upon to assess the merits of his own work or that of a competitor.

According to figures produced by Armstrong and statements made in the House of Commons, the charge of extravagance brought against the authorities was equally groundless. Elswick-made guns did not cost £2,000 apiece, or anything like that sum. Armstrong's first experimental gun had cost £1,000. The prices charged for some 1,600 guns completed between 1856 and 1861 ranged from £120 to £650 or more, according to size.

As for the Elswick-made guns sent to China, the Commander-in-Chief of the British Expeditionary Force spoke highly of them. An officer cited in a popular magazine as authority for the statement that some of them had to be withdrawn because they were unreliable publicly denied that anything of the kind had happened. He added that, on the contrary, the reputation of Elswick-made guns was such that they were always the first to be sent forward when artillery support was called for.[7] Armstrong claimed, on the strength of official returns, that only twenty of the guns made at Elswick up to 1861 had had to be returned to the factory for repair, and that not a single one of them had burst.[8]

The fact remains that a good many adverse comments on the new-fangled breech-loaders were made by users inclined to judge them more by what crews accustomed to muzzle-loaders could do with them in the field than by what was said about them by expert witnesses or Members of Parliament. After hearing a great deal of evidence and discussing it for months on end, the committee of enquiry succeeded, by a process of hard bargaining between its members, in producing a report described long afterwards by Stuart Rendel as 'fairly capable of construction as generally favourable to Armstrong'. And he did not disguise his conviction that, but for his efforts, the verdict might have gone the other way.[9]

In any case, long before the committee submitted its report, and even before it began its deliberations, the authorities were heartily tired of the stream of censure to which their sponsorship of the Elswick Ordnance Works exposed them. No doubt partly to escape further criticism, but chiefly to save money, Lord Palmerston's government exercised in 1862 its right to terminate the Elswick contracts. The reason given at the time was that henceforth the output of a single factory would meet the country's needs.

This did not prevent the authorities from toying for a time with a proposal that guns should be ordered from Krupp. Nevertheless they did have some reason to suppose that possibly the best answer to their problems of procurement and supply might be to make themselves independent of such undertakings as the Elswick Ordnance Company by increasing the output of ordnance at Woolwich. Some years earlier the failure of commercial contractors to make punctual delivery of a large batch of Minié rifles had led to the despatch of a mission to the United States to study what was called 'the American system' of small-arms production. The mission visited Springfield National Armory and a number of commercial small-arms factories. In the light of the information it brought back, mechanized plant was installed at the government small-arms factory at Enfield Lock and the post of Superintendent was offered to a citizen of the United States, James Burton of Harpers Ferry National Armory. Manufacture by mass-production methods of the 1853 Enfield rifle—an improved Minié rifle designed in that year but introduced in 1855—began in 1858. The new system proved such an advance on the old sub-contracting method that output soon reached well over a thousand rifles a week.

That Krupp ranked by the 1860s as a possible supplier of ordnance to the British government may seem surprising in view of his failure to obtain a single order for guns when he showed his famous all-steel cannon at the Great Exhibition in 1851. Apart from selling some guns to the Khedive of Egypt and one to the Tsar of Russia he had, in fact, achieved so little success as a gunfounder for some years after that experience that by 1859 he was thinking of selling or scrapping his gunmaking plant and giving up the whole idea of making armaments a profitable side-line.[10] But in that year a decision by the Regent Wilhelm of Prussia to modernize the Prussian Army brought him an order for more than three hundred guns. The outlook for the future Cannon King became still brighter when Wilhelm made up his mind, after succeeding to the throne, to defy Liberal and pacifist critics by summoning the reactionary Otto von Bismarck to office as Minister-President. Bismarck took up his post in 1862 with a mandate to carry through the reorganization of the army at all costs and an unshakeable determination to prepare in secret for war with Austria.

The result was that a steady demand from the Prussian military authorities for rifled cannon enabled Krupp to build up a flourishing export trade in these weapons. For some years the Russians were his best customers outside his own country. But he also did business with Brussels, The Hague, Madrid, Berne and—in defiance of a strong hint from Wilhelm—Vienna. To England he exported not guns, but inner barrels used by Armstrong for his built-up breech-loaders. He seems, however, to have counted these as guns when compiling a list of his foreign markets.

Soon after the termination of the Elswick contracts, the Elswick Ordnance Company ceased to exist as an undertaking distinct from Armstrong's engineering firm. Armstrong's appointment as Engineer of Rifled Ordnance still had a year or two to run, but he felt it would not become him to go on drawing his salary while his associates at Elswick faced an empty order book. He resigned his post and returned to Newcastle to become senior partner in Sir W. G. Armstrong and Company, a company formed for the purpose of uniting the two sides of the business. The other partners were George Cruddas, his son William Cruddas, Richard Lambert, George Rendel, Andrew Noble, and the manager of the engineering works, Percy Westmacott.

As a result of this transaction the new company acquired, among other assets, an arms factory employing about 3,000 men and equipped at a capital cost of £168,000.[11] Painfully aware that there was no longer a market in Britain for the fifty tons of guns a week the factory was capable of producing, the partners offered the plant and machinery to the government at their written-down value of £137,000. The government declined this proposal but agreed to pay compensation for cancelled orders on a scale to be determined by independent assessors. Eventually the company received £65,000 in cash and was allowed to keep plant worth about £20,000 whose purchase had been financed from public funds.

The authorities claimed after the termination of the Elswick contracts that half a million pounds had been spent at Elswick on guns which would have cost half as much had they been made at Woolwich. This assertion was sharply challenged at the time and remains questionable to this day. Since the value of the orders placed by the government with the Elswick Ordnance Company between 1859 and its demise in 1863 was not half a million pounds but more than twice that sum,[12] presumably what the authorities meant was that roughly half the guns made at Elswick could have been made at Woolwich and that they would then have cost not half a million pounds but roughly a quarter of a million. To offer any conclusive proof of either of these statements seems to have been more than the authorities could manage in 1863, and would be utterly impossible after the lapse of more than a century.

Having failed to sell their gun-making plant and machinery to the government, Armstrong and his associates had to choose between closing the arms factory and finding new markets for its products. Closing the arms factory would have meant writing off a substantial part of the new company's capital and sacking three-quarters of its entire labour force, at any rate until new jobs could be created at the engineering works. Not surprisingly, the partners decided against so drastic a measure. Since the government was the only purchaser of arms in Britain, the logical and inevitable consequence of their decision was that Sir W. G. Armstrong and Company entered the international arms trade in competition with Krupp and any other arms manufacturers who might be on the lookout for foreign business.

At first, Armstrong was unwilling that foreign buyers should be approached so soon after he had given up a government post. Stuart Rendel told him he could not afford to wait. Krupp had already gained a foothold in the European market and in Egypt. Joseph Whitworth had begun to do business in South America.

Before the termination of the Elswick contracts, the Brazilians had wished to place an order for Elswick-made guns. They had turned to Whitworth when they found the Elswick Ordnance Company was still bound by its undertaking to sell only to the British government. Nothing, said Rendel, could be worse for the new company than that foreign governments should seem to prefer Whitworth's guns to Armstrong's. If Armstrong did not seize his opportunities while there was still time, newcomers to the arms trade would supplant him. Rendel added that, in his opinion, there was nothing unpatriotic about selling arms abroad. On the contrary, it would tend to enhance Britain's power and influence by making foreign countries dependent on her for the means of defence.

In the light of these arguments, Armstrong agreed that Rendel should be at liberty to travel abroad on the company's behalf and pick up such business as he could. 'To make it worth your while,' he wrote, 'we will give you five per cent commission upon the orders you bring us.'[13]

This was a much higher rate than afterwards became usual in the arms trade. But neither party had any cause to repent of the bargain. Well-mannered, persuasive, keenly aware of the importance of not frightening prospective purchasers away by pressing them too hard, Stuart Rendel proved an excellent salesman. Armstrong's reputation stood so high that the company could afford to charge a good price for its products. Within a few years its customers for guns included the governments of Austria, Chile, Denmark, Egypt, Italy, the Netherlands, Peru, Spain and Turkey. It also sold arms to both sides in the American Civil War. The British government ordered a few guns from time to time, but the total value of these orders did not amount during the first fifteen years of the new company's existence to more than £60,000 or so.

In most of these transactions Rendel played an important part. The Italians placed their first order for naval guns in the spring of 1864, took delivery of arms to the value of £100,000 in 1866, and soon adopted Armstrong guns as the standard equipment of their warships. For some years all their business with the company was transacted through Stuart Rendel and their naval attaché, at that time a captain living in humble lodgings south of the Thames but afterwards an admiral and resident director of an Armstrong subsidiary in Italy. The Prussians could not be weaned from their allegiance to Krupp, but the Egyptians were persuaded to buy Armstrong guns after they had tried Whitworth's as well as Krupp's. In his negotiations with the Khedive, Rendel used as intermediary his friend George Joachim Goschen, a future First Lord of the Admiralty and Chancellor of the Exchequer. The Austrians remained faithful to Krupp so far as weapons for their army were concerned, but they did buy from Armstrongs naval guns, including the armament of two large warships. The Russians bought an experimental muzzle-loader and offered to abandon Krupp and give up their own armament factory at Alexandrovsky if Armstrongs would establish a subsidiary company in Russia; but the matter was not pursued, perhaps because Rendel was warned that doing business on any considerable scale with Russian ministers and officials would involve him in transactions of a kind a budding British statesman could not afford to countenance.

However, Rendel was not content to be merely a salesman. Before long he was both the company's *de facto* London manager and a partner. He began with a one-twenty-fifth share bought from Armstrong for £19,600. On Armstrong's death in 1900, he became the largest individual shareholder in the multi-million-pound company formed a few years earlier to amalgamate the Armstrong and Whitworth interests.

Within a few years of its formation, Sir W. G. Armstrong and Company created a new outlet for its products by arranging that a Tyneside shipbuilding firm, Charles Mitchell and Company of Low Walker, should build a number of gunboats specially designed to carry Armstrong guns. About twenty of these small, unarmoured vessels were built during the next fifteen to twenty years and sold to the Admiralty or to foreign governments. Armstrongs went on to become warship constructors as well as manufacturers of naval guns by taking over the Low Walker yard in 1882 and laying down a fast cruiser, the *Esmeralda*, for the Chilean government. Described by Armstrong as the swiftest and most powerfully armed cruiser in the world, the *Esmeralda* had a maximum speed of slightly more than eighteen knots and carried two 10-inch and six 6-inch guns. She was unarmoured, but her engines, boilers and magazines were all below the waterline and protected by a watertight steel deck. Sold after some years of service with the Chilean Navy to the Japanese and renamed the *Idzumi*, she became famous as the ship from which the approaching Russian fleet was first sighted before the decisive battle of Tsushima in 1905.

Some twelve months after the *Esmeralda* was laid down, Armstrongs and Mitchells joined forces. Sir W. G. Armstrong, Mitchell and Company was formed in 1883 as a limited liability company with a nominal capital of two million pounds to unite the two concerns under Armstrong's chairmanship. A warship building yard was established in the following year at Elswick. Thereafter the Low Walker yard built only merchant vessels.

In the meantime Armstrongs became manufacturers of other weapons besides rifled cannon. In 1870 they entered into an arrangement with the Gatling Gun Company which made them sole producers in Britain of the Gatling gun, invented by Richard Jordan Gatling of Hertford, North Carolina. This was a multi-barrelled semi-automatic weapon with a calibre of half an inch. It fired up to 350 rounds a minute when a handle was turned. Later, the rate of fire was increased to 400 to 500 rounds a minute and customers could choose between ·50-inch and ·65-inch models which sold in Britain at £205 and £265 respectively. Armstrongs were paid rather more than half the selling price of each gun they manufactured for the company, and from 1873 they also received a commission of five per cent of the price of every gun they sold.

Stuart Rendel succeeded between 1870 and 1877 in selling a satisfactory number of Gatling guns to the British government for use in warships. In 1878, however, he and Edgar Welles, Secretary of the Gatling Gun Company, learned to their dismay that the Admiralty had formed a favourable opinion of a rival weapon. This was the semi-automatic gun invented by Thorsten Nordenfelt, the Swedish engineer who had the distinction of being the first arms manufacturer to give employment to the notorious Basil Zaharoff. Later in the year, the authorities decided to hold comparative trials of four different semi-automatic weapons.

The outcome of the trials was that a government committee reported in favour not of the Gatling or the Nordenfelt gun but of a relatively obscure competitor whose chief merit was its simplicity. Nevertheless the Nordenfelt gun sold so well in a number of countries that in 1886 a British company, the Nordenfelt Guns and Ammunition Company, was established under the chairmanship of Admiral Sir Astley Cooper Key, a gunnery expert, to exploit Nordenfelt's patents. A site for a factory at Erith, in Kent, was acquired in the following year.

By that time Nordenfelt had become interested in the development of steam-driven, torpedo-carrying submarines based essentially on the work of a Liverpool clergyman, the Reverend G. W. Garrett. In 1881 he took out a patent of his own for 'submarine or subaqueous boats or vessels'. Five years later he arranged that two submarines should be built for him by the Barrow Shipbuilding Company, which had come into existence in 1871 and in 1877 had begun building warships for the Admiralty. The Nordenfelt submarine was destined to be superseded by submarines developed from a design by an Irishman, J. D. Holland, but a consequence of Nordenfelt's deal with the Barrow Shipbuilding Company which is not without historical interest is that the company received permission in 1887 to call on the services of foreign agents employed by his Swedish arms company. The most notable of these was Zaharoff, at that time acting as Nordenfelt's agent or correspondent in the Near East.

The torpedoes carried by Nordenfelt's submarines and used by the Royal Navy from 1871 were not the underwater mines called by that name in earlier times. From 1866 the term was applied to the self-propelled missile invented in that year by Robert Whitehead, a British engineer domiciled at Fiume. Whitehead offered his invention to the British government, but the Admiralty turned it down on the ground that no good could come of fostering a weapon that might be used to destroy Britain's command of the sea. He then set up a factory at Fiume. His firm, in which his son John Whitehead and his son-in-law Count George Hoyos became partners, traded as Whitehead and Company until 1905. It was then incorporated under Hungarian law as a limited liability company with the English title of Whitehead and Company Limited and titles in three other languages. On learning at the beginning of the 1890s that the Admiralty did not propose to give them any more orders unless they opened a factory in Britain, the partners established an English branch of the business at Weymouth.

The Whitehead torpedo and a new generation of fully-automatic weapons destined to succeed the Gatling and the

Nordenfelt gun were to have profound effects on twentieth-century warfare. They were to influence not only strategy and tactics but also the design of weapons, equipment and ammunition and the scale on which weapons, equipment and ammunition would be demanded by twentieth-century belligerents. The advent of the Nordenfelt gun was also a portent for arms manufacturers in another sense. It was a sign that competition in the arms trade was becoming tougher, and indeed was to become tougher with every year that passed. The age of mergers, of determined and often successful attempts by arms manufacturers to increase their share of the market by devouring their rivals, was close at hand.

FOUR

The Armament Kings

Soon after the Crimean War, the French industrial magnate Eugène Schneider became convinced that he was threatened with ruin by a new commercial treaty between France and Britain. Fearing that he might lose his dominance of the home market unless he acted promptly, he hastened to diversify his output and improve its quality. The company did not open a separate gun-founding department until 1888. But from 1867, when it entered the steel industry, its products included steel ingots and forgings for gunmaking, and it played a leading part in the re-equipment of the French Army after the Franco-Prussian War.

Following a tradition which went back to the Crimean War, the company also took a keen interest in the development of armour for warships. Eugène Schneider died in 1875 and was succeeded by his son Henri. In the following year Henri sent to Spezia, for testing by the Italian Navy, armour whose acceptance by the Italians gave the company a prominent position among suppliers of armour plate to the world's navies.

In the meantime Bismarck's determination to oust Austria from her leadership of the German states culminated in the war for which he had long been preparing. On 3 July 1866, the Prussians inflicted a crushing defeat on the Austrians near the villages of Königgrätz and Sadowa, in Bohemia. Their victory was widely attributed at the time to the excellence of their equipment, but this was only part of the story. Their troops, armed with breech-loading rifles based on Nicholas Dreyse's 28-year-old design, were able to fire much more rapidly than the Austrians with their muzzle-loaders. But Krupp's breech-loading cannon were far from satisfactory. So many of them blew up or proved otherwise defective that he was driven to the verge of despair and was able to retain the goodwill of the authorities only by promising to replace all faulty weapons at his own expense.

The next item on Bismarck's agenda was war with France. In the summer of 1870 he succeeded, by publishing a carefully edited account of an interview between Wilhelm and the French Ambassador, in goading the French into a declaration of war.

As in 1866, the Prussians were ready and their enemies were not. Three German armies marched into France while the French were still mobilizing their forces for a projected advance into southern Germany. A series of victories gained by the Prussians and their allies on French soil culminated in the surrender of more than a hundred thousand French troops at Sedan and a still larger number at Metz. The French monarchy fell, Paris was invested, and Wilhelm was proclaimed Emperor of Germany in the Hall of Mirrors at Versailles. The peace treaty condemned France to pay a huge indemnity and cede the greater part of Alsace and Lorraine to the new German Empire.

This time Krupp's guns acquitted themselves well. Their range and reliability still left much to be desired, but the accuracy of the Prussian artillery fire at Sedan astonished all observers and was largely responsible for the inability of the French to break out of their positions once they were encircled. One consequence was that the Chinese, hitherto accustomed to buying their guns from Armstrongs, placed a substantial order with Krupp.

After the war, the French held comparative trials at Bourges of a number of guns used by European armies. One of the conclusions drawn from them was that the British Army's Woolwich-made rifled muzzle-loader was as good a weapon as any of those tested. But Armstrongs developed soon afterwards a much improved breech-loading 6-inch gun which Armstrong himself considered the most important innovation of its kind since the introduction of rifled ordnance. In 1878 strained relations with Russia led the British government to requisition some guns intended by Armstrongs for delivery to Italy. The new 6-inch guns made such a favourable impression that the military authorities discarded almost overnight the prejudice against breech-loaders that had grown up since the termination of the Elswick contracts. In the same year Armstrongs received the first large order for guns for the British Army that had come their way for nearly fifteen years. Whitworths, too, received a substantial order.

Soon afterwards the Admiralty, newly empowered to choose its own guns instead of obtaining them through the Ordnance Department, overcame a distrust of breech-loaders which had been even stronger in the navy than among soldiers. From that time until the outbreak of the First World War, Armstrongs and their principal competitors in the domestic market received a stream of orders from the Admiralty for naval guns and gun-mountings, but only occasional orders from the Ordnance Department, which continued to rely on government arsenals for roughly four-fifths of the army's needs. It was the business they did with foreign governments and the Admiralty, not the patronage of the Ordnance Department, that enabled British arms manufacturers to keep in the forefront of technical progress by spending substantial sums on research, development and design.

Meanwhile the widespread adoption of the Whitehead torpedo confronted the world's leading naval power with formidable problems. In the 1870s a British firm, John I. Thornycroft and Company, took the lead in introducing small fast vessels similar to gunboats but armed with torpedoes. These short-range mini-warships posed such a serious threat to capital ships and cruisers that a new kind of warship, the torpedo-boat destroyer, had to be developed to counter them. But the destroyer was essentially a fast ship with a limited radius of action. In the days of sail the British had been able to blockade an enemy's ports for months or even years on end by sending warships to watch them. To do this with coal-burning or oil-fired steamships which had to be refuelled at frequent intervals would be far more difficult, but the Lords of the Admiralty continued to hope until well into the twentieth century that it would still be possible. They were then forced to admit that the advent of the torpedo boat, to say nothing of the torpedo-carrying submarine, had made close blockade impractical. So many destroyers, working in relays, would be needed to protect cruisers or capital ships operating for long periods off a hostile coast that not even the richest country in the world could afford to build enough of them. When war with Germany became a distinct possibility, the Admiralty had to fall back on a system of distant blockade designed to hamper the enemy's foreign trade and make it difficult for his major warships to gain access to the High Seas without detection.

Naval strategy and tactics were also profoundly affected between the 1870s and the outbreak of the First World War by the progress of naval gunnery. So, too, were the foreign policies of some first-class powers. Improvements in the range and accuracy of guns and the penetrating power of projectiles brought demands for stronger armour, whose introduction in turn brought demands for even more effective guns. Capital ships became so expensive, the struggle for naval supremacy so gruelling, that eventually Britain abandoned her policy of maintaining a battle fleet comparable with the combined battle fleets of her two strongest rivals and defied race prejudice by making an alliance with Japan. According to Lord Salisbury's government of 1895 to 1902, the alternative to abandonment of the two-power standard would have been to allow the annual cost of the navy to rise within the next six years to ten times the average of the past decade.

The naval arms race created wonderful opportunities for shipbuilders, arms manufacturers and suppliers of armour at a time when British heavy industry was feeling the effects of American and German competition. Armstrongs, as we have seen, became warship constructors as early as 1882, when they laid down the *Esmeralda*. In 1888 the Naval Construction and Armaments Company, in which Thorsten Nordenfelt had an interest, was formed to take over the business of the Barrow Shipbuilding Company at Barrow-in-Furness. In the same year Charles Cammell and Company, the well-known Sheffield steel firm, submitted for test at Portsmouth the first successful all-steel armour made in Britain. Vickers, Sons and Company, also of Sheffield, submitted soon afterwards solid steel armour, ten and a half inches thick, which the naval authorities found so satisfactory that they placed a large order for it. Within a few years Vickers were manufacturing armour in substantial quantities and were employing one of Nordenfelt's agents, Basil Zaharoff, to sell it to foreign governments.[1]

However, in 1891 the United States government tested, with satisfactory results, all-steel armour made by a process invented by Hayward Augustus Harvey of New Jersey. This was manufactured in the United States by the Bethlehem Iron Company and Carnegie, Phipps and Company. Harvey took out a British patent for his process, but just before he did so a Sheffield steel firm, John Brown and Company, introduced a method almost identical with his. John Brown's method was taken up by Charles Cammell and Company and Vickers, Sons and Company. The result was that for some time all three of the leading Sheffield steel firms manufactured armour which closely resembled that made by the Harvey process, but paid no royalties to Harvey.

Eventually all three firms acquired licences from Harvey. But by that time he had set up, in addition to a parent company in the United States, a number of associated companies whose interests tended to conflict with those of licensees, and which sometimes found themselves competing with each other. In Britain, for example, there were two Harvey companies and three licensees. In the United States there were the Bethlehem Iron Company and Carnegie, Phipps and Company as well as Harvey's parent company. There was also a Harvey company in France.

To put an end to this unsatisfactory state of affairs, the four Harvey companies agreed in 1894 to form a syndicate for the purpose of sharing markets and fixing prices, and to invite other manufacturers to join them. The outcome was an arrangement by which the world's leading producers of armour plate agreed to divide up orders from 'outside' countries by drawing lots. This did not apply, however, to orders from the British, French, German and United States governments. In addition to the Harvey companies, the parties to the agreement were, in the United States, the Bethlehem Iron Company and Carnegie, Phipps and Company; in Britain, John Brown and Company, joined later by Armstrongs; in France, Schneider et Compagnie, the Aciéries de la Marine and the Compagnie des Forges de Chatillon; in Germany, Alfried Krupp's son and heir Friedrich Krupp and the Dillenger, Hütte Werke.

Here was just such a price-fixing, market-sharing agreement as might seem, at first sight, to justify the allegation made in 1921 that arms manufacturers had conspired to increase the price of armaments sold to governments. However, so far as Britain, France, Germany and the United States were concerned, clearly the agreement did not have that effect, for armour sold to the governments of those countries was expressly excluded from its scope. The only governments that could have been adversely affected by it were those of 'outside' countries, and to these the members of the syndicate owed no allegiance.

Within a few years of the formation of the Harvey Syndicate, Krupp introduced a process which yielded a nickel-chromium steel plate of high quality. By 1897 all the leading British firms held licences from Krupp as well as Harvey. From that time until the Krupp patents expired some thirteen to fourteen years later, the Admiralty invariably specified armour made by the Krupp process. The result was that for some years Krupp received a royalty of £4 or £5 on every ton of armour used in British warships. This was a consequence not of any conspiracy among arms manufacturers to raise prices, but of the Admiralty's preference for a particular kind of armour which, as the authorities were well aware, could only be made under licence from a German firm.

What, then, were the 'international armament rings' and 'international armament trusts' supposed by critics of the arms trade to have accentuated the armament race and raised the price of armaments sold to governments? The Steel Manufacturers' Nickel Syndicate was a limited liability company formed in 1901 for the purpose not of raising prices but of ensuring that British users of nickel received fair treatment from the French company which controlled the nickel deposits of New Caledonia, at that time the sole source of supply. The Union des Mines was an association of German and French steel magnates interested in exploiting the mineral resources of Morocco. Because France and Germany nearly came to blows about Moroccan questions, it has been widely assumed that the Union exerted a malign influence on Franco-German relations. In fact, a confrontation was the last thing its members wanted. Eugène Schneider, son and successor of Henri Schneider and grandson of the founder of the firm, withdrew from the Union when the Agadir incident of 1911 showed that the statesmen could not be relied upon to avoid one.

This does not mean, of course, that rich and powerful armament firms such as Krupp and Schneider exerted no influence at all on public affairs. Their existence was itself a factor in international politics. The fact that Germany possessed an arms industry with an unsurpassed war potential must undoubtedly have weighed with the Kaiser and his Foreign Minister when they decided in 1905 to test the strength of the new understanding between France and Britain by launching a diplomatic offensive in Morocco. No doubt they were anxious, too, that the Sultan should not be persuaded by the French to discriminate against German goods. But it is not to be supposed that the Kaiser was despatched to Morocco with a mandate from Essen to warn his host that Germany expected him to buy his arms from Krupp, not Schneider. Three years before the visit was made, Friedrich Krupp had died in mysterious circumstances at a moment when he was implicated in a resounding scandal arising from homosexual frolics to which he had become addicted while yachting in Italian waters. The family firm had been transformed into a company in which his elder daughter Bertha held all but a handful of the shares. Bertha was still in her teens. It is hard to imagine the Kaiser receiving a lecture on foreign policy from the civil servants, lawyers and technical experts who played the chief part in directing the company's affairs during her minority.

In 1906 Bertha married Gustav von Bohlen und Halbach, thereafter known at the Kaiser's behest as Gustav Krupp von Bohlen und Halbach. A descendant of two German-American families, the bridegroom had been serving in a diplomatic post when he first met Bertha. Middle-aged, a pillar of respectability but stolidly ambitious, he assumed control of Friedrich Krupp AG in the capacity of steward and trustee of his wife's fortune. Tilo von Wilmowsky, an East Prussian landowner of liberal views who had married Bertha's younger sister Barbara, agreed to join the board because, he said, he felt Gustav needed all the help he could get.

In the second half of the nineteenth century, Alfried Krupp had invested hugely in sources of raw material for his products. Friedrich—known to his family as Fritz—had added a shipyard, a controlling share in a company which made high-quality armour plate, and an interest in ballistite, the smokeless powder invented by Bernhard Nobel. Gustav branched out into the manufacture of barbed wire, used in the Russo-Japanese war for the protection of field works. By the second decade of the twentieth century the firm was the world's largest producer of diesel engines for submarines. Its assets included sixteen factories or steel plants at Essen or elsewhere in Germany, three artillery ranges, and foundries, mines and quarries as far away as Silesia. On the eve of the First World War Bertha Krupp was reckoned easily the richest of the Kaiser's subjects, with an annual income estimated at roughly a million and a quarter pounds or approximately six million dollars.

Eugène Schneider's interests were equally far-reaching and diverse. In 1897 his father had acquired the ordnance department of the Société des Forges et Chantiers de la Méditerranée. This included an up-to-date arms factory established at Le Havre in the 1880s under the supervision of the celebrated artillerist Gustave Canet, and an artillery range suitable for trials of the largest guns. Other assets inherited by Eugène in 1898 or acquired during the next year or two included a ship- and bridge-building yard on the site of the old marine department at Chalon-sur-Saône, ironfields in Spain, and blast furnaces, steel works, rolling mills, electrical and engineering works and coal mines and iron-fields in various parts of France. Eugène also held large blocks of shares in a variety of French and foreign mineral and industrial undertakings.

Despite the popular belief that Krupp's chief customer for arms was the German government, the position in 1911 was that more than half the firm's output of guns up to that time had been sold abroad. In the Balkan wars that preceded the First World War, the Greeks and the Bulgarians were armed chiefly with guns supplied by Schneider, the Turks and the Rumanians with guns bought from Krupp. It is natural to assume that rivalry between the two firms played some part in exacerbating mutual distrust and hostility between France and Germany, but impossible to point to any stage at which their influence on national policies was or could have been decisive. Gustav and Bertha Krupp and Tilo von Wilmowsky were in no way accountable for the plan for the invasion of France by way of Holland and Belgium drawn up by Alfred von Schlieffen at the time of the first Moroccan crisis. Nor were they responsible for the plan to invade France by way of Belgium and Luxembourg initiated by Helmuth von Moltke about the time of Agadir. The firm helped to make Moltke's plan feasible by producing specially-designed giant howitzers used in 1914 to reduce the Belgian forts at Liège; in addition, Moltke borrowed from the Austrians some large howitzers, or 'mortars', made in the arms factory established by Emil von Skoda at Pilsen in Bohemia. But the existence of close links between the firm's technical experts and the military authorities does not necessarily imply, any more than Gustav's and Bertha's friendship with the imperial family necessarily implies, that anyone connected with the firm was consulted about matters of high policy. Had Gustav been privy to the secrets of the War Ministry, his henchmen would scarcely have needed to bribe officials to purloin documents on their behalf, as two of them—one a director of the firm—were convicted of doing not long before the outbreak of the First World War.

As for General Joffre's plan for the defence of France in the event of a German invasion, it was deemed so secret that not even his army commanders or the British were allowed to see it in its entirety. It may be thought that, in a country as corrupt as Republican France, a man as rich as Eugène Schneider was unlikely to be without influence in high places. But he cannot have had much pull in military circles. Had his influence been decisive, the Briey ironfields would scarcely have been so inadequately defended that the Germans were able to capture them soon after the outbreak of war. Nor would the French Army, one may think, have ordered so few of the medium and heavy guns his company could so well have made.

In fairness to Moltke and the Kaiser, it must also be remembered that the war that erupted in 1914 was not precisely the war for which they had been preparing for some years. Moltke did not foresee—how could he have foreseen?—that Europe would be convulsed by the outcome of a quarrel between Austria and Serbia, arising from a crime not yet committed when his plan was drafted. Moreover, his paper scheme made very inadequate provision for British intervention. As for the Kaiser, he was so appalled by the prospect of simultaneous war with Russia, France and Britain that he reduced Moltke to 'tears of abject despair' by trying to stop the invasion of Luxembourg after the mobilization order had been issued. His great mistake was that he failed to exert a restraining influence on the Austrian government while there was still time to avert a major conflict. And for that error no arms manufacturer can be held responsible. The Kaiser was not a man who would willingly have listened to counsels of moderation from representatives of Krupp.

In England, as in Germany, there was a certain amount of lobbying before the war, both by members of the general public and by interested parties, in favour of a strong navy. Ministers were accused from time to time of showing undue favour to commercial firms, not necessarily connected with the arms trade, but on the whole the attitude of British governments to leaders of commerce, industry and finance for many years before 1914 was extremely cautious. Not long before the outbreak of war the Treasury gave a guarantee to the Ottoman Bank and the state acquired a controlling interest in the Anglo-Persian Oil Company; but these were exceptional measures, held to be justified by the need to check German economic penetration of Turkey and ensure a supply of oil fuel for the navy. Representatives of certain branches of commerce and industry were consulted during the preparation of the Government War Book, which made provision for steps to be taken on the outbreak of war or the declaration of a state of national emergency. But so far as the arms trade was concerned, the general policy was to restrict contact between commercial firms and the service departments to technical discussions at official level. Armstrong was politely rebuffed when, in 1870, he suggested a meeting with the First Lord of the Admiralty to discuss questions affecting the armament of the fleet. Even in 1914, when the leading armament firms were regarded as 'partners with the Admiralty in a common task', their business was transacted, as a rule, not at the ministerial level but with the Third Sea Lord, the Director of Naval Construction and other senior officers or officials and their staffs.

By that time some of the leading British armament firms were linked with foreign arms manufacturers, and in some cases with each other, by complex licensing agreements or other special arrangements. To understand how some of the most important of these came into existence, it is necessary to go at least as far back as 1881.

In that year Hiram Maxim, an American of Huguenot descent whose family had settled for a time in England before emigrating to the United States, attended an international exhibition in Paris on behalf of an American electrical firm. A man of many talents, Maxim was a skilled carpenter, metalworker, engineer and draughtsman. He was also an accomplished boxer, wrestler and teller of tall tales. He did not smoke or drink beer, wine, brandy, whisky, gin or coffee, but enjoyed fighting and gambling. While in Paris he roughed out the design of a fully automatic machine-gun which utilized the recoil from the firing of a round to load and fire the next round.

This proved an epoch-making weapon. It became the ancestor of a whole family of automatic weapons which helped to make prepared positions well-nigh proof against orthodox infantry assaults unless they were first reduced by exceptionally severe or prolonged artillery bombardments. Thus the ultimate effects of Maxim's invention were to give a powerful stimulus to the design and production of field artillery and to create unprecedented demands for ammunition suitable for knocking out machine-gun posts and countering the enemy's supporting fire. These trends were accentuated by improvements in the design of small arms and standards of musketry, and by a growing awareness of the value of entrenchments and barbed-wire entanglements in the light of experience in the South African, Russo-Japanese and Balkan wars.

From Paris Maxim travelled to London, where he completed the detailed drawings for his design in a rented office. He had then to find a firm able and willing to manufacture and market his brain-child on acceptable terms. Probably he would have offered his invention to Armstrongs had they not been committed to the Gatling. As things were, he decided to approach Vickers, Sons and Company.

Vickers, Sons and Company was a family firm with its roots in Yorkshire but linked by powerful ties with the United States. In 1829 William Vickers, with George Portus Naylor and others, had founded Naylor, Hutchinson, Vickers and Company—from the 1840s called Naylor, Vickers and Company—as an offshoot of the old-established Sheffield firm of Naylor and Sanderson. The Vickers of Vickers, Sons and Company was Edward Vickers, a brother of William Vickers and married to a sister of George Portus Naylor. The sons were Edward's sons.

Edward Vickers had begun his commercial career as a miller. But he was largely responsible for his brother's decision to enter the steel industry, and had taken a leading part in the management of the business almost from the start. He was the first steel manufacturer in England to make cast tyres and heavy forgings by a method introduced by Jacob Mayer of Bochum, and had sent two of his sons, Tom and Albert, to Germany to receive part of their technical education in that country. In 1867 financial stringency resulting from a series of bank failures had compelled him to turn the firm into a limited company in which half the shares were held by his New York agent and former partner, Ernst Leopold Schlesinger Benzon. This infusion of capital proved so beneficial that by 1881, when a record profit was earned, the company's assets were worth at least four times as much as in 1867. Benzon had died, immensely rich, in 1873, and Tom Vickers—known to his employees as Colonel T. E. Vickers—had succeeded him as Chairman. Edward Vickers was still alive and still keenly interested in the company's fortunes.

When Maxim approached them in 1882, Vickers, Sons and Company were not yet arms manufacturers, although they had supplied steel for the making of ordnance and held a patent for a method of annealing and cooling blocks of iron and steel for gun-making. They did not, in their corporate capacity, take up Maxim's offer. A separate company, the Maxim Gun Company, was established as a limited liability company in 1884 to exploit Maxim's invention. Members of the Vickers family acquired a substantial though not a controlling interest, and Tom's brother Albert became the new company's first Chairman. Maxim decided to stay in England and eventually became a British subject and was knighted. A factory was built at Crayford, in Kent, and in 1885 a new, large-calibre version of the Maxim gun was introduced. The new gun, Albert told his father, was such an 'absolute success' that he would not sell it outright for less than a quarter of a million pounds.[2]

However, the superiority of the Maxim gun to rival weapons cannot have seemed obvious to Thorsten Nordenfelt and his English associates. Encouraged by the success of the Swedish-made version of Nordenfelt's gun, they went ahead with the building of their new factory at Erith, only a few miles from Crayford. Both guns found buyers in a number of countries, but competition between the two British companies became so keen that it might have proved ruinous to both had not two astute financiers, Lord Rothschild and Ernest Cassel, succeeded in 1888 in uniting them as the Maxim-Nordenfelt Guns and Ammunition Company. Rothschild and Cassel acquired, between them, approximately as many shares in the new company as were allotted to members of the Vickers family. Zaharoff became an agent for the Maxim-Nordenfelt company, and in 1890 was appointed the company's 'foreign adviser' at a fee fixed in the following year at 'not less than £1,000 per annum'.[3]

We shall see later that various accounts have been given of the way in which Zaharoff came to know Tom and Albert Vickers and to work for the family firm. But it does not seem necessary to look far for an explanation of his appointment as a foreign agent for the armour first made by Vickers, Sons and Company in 1888. His work for the Maxim-Nordenfelt company, and Albert's association with that company, would make Zaharoff an obvious candidate.

In the year in which they became manufacturers of armour, Vickers, Sons and Company entered the arms trade by another door. In 1884 government emissaries had visited three British steel firms, of which Vickers was one, for the purpose of assessing their potential as manufacturers of ordnance. All three firms said later that they had been induced by these visits to enlarge their works in anticipation of orders which, so far as two of the firms were concerned, they did not receive. But Vickers had begun to enlarge their works before the visit was made, and their foresight was rewarded, albeit after a long delay. In 1888 they received an order for guns and set up a special department to deal with it. In the summer of 1890 their first large gun passed its tests at Woolwich.

However, orders for guns received from the British government during the next few years were not nearly enough to give full employment to the new ordnance department. Orders for large guns, in particular, were few and far between. Moreover, for some years business in general was bad. Profits between 1887 and 1893 were so much smaller than those earned between 1881 and 1886 that ordinary shareholders received only meagre returns after dividends had been paid on preference shares issued to finance the company's entry to the arms trade. They picked up in 1894, but even then shareholders were little better off than they had been before the ordnance department was created. The directors, who now included Tom's son Douglas Vickers, came to the conclusion that the company must face the logical consequences of the steps it had taken in the past few years. In order to profit by a worldwide demand for naval armaments, it must equip itself to build warships complete with their engines, armament and armour. It must become not so much a steel firm with a substantial interest in the arms trade as an armament firm with a substantial interest in steel.

Accordingly, in 1897 the company bought out the Naval Construction and Armaments Company at Barrow. It thus acquired, at a cost of £425,000, docks covering 294 acres on which £2,225,000 had been spent and at which five thousand workmen were employed. Three cruisers had been completed at Barrow in recent months, and a fourth cruiser and four destroyers were under construction there.

Like Vickers, Sons and Company, the Maxim-Nordenfelt Guns and Ammunition Company had its ups and downs in the early 1890s. The company did quite well for some years after the merger of 1888. But in 1894 it made a loss and gave Zaharoff six months' notice, although his appointment was renewed soon afterwards on terms which promised him a commission of one per cent on all orders from Continental Europe. In the following year Lord Rothschild brought about the appointment as full-time Managing Director of Sigmund Loewe, a friend of Rothschild's London manager and already, at Rothschild's prompting, a member of the board. Loewe gave up his job as manager of a small explosives company to devote himself so successfully to the affairs of Maxim-Nordenfelt that the company, after losing money for two years in succession, made a profit of £138,000 in his first full year of office.

Sigmund Loewe had a German-domiciled brother, Ludwig Loewe, who was until 1896 a manufacturer of sewing-machines. In that year Ludwig founded the Deutsche Waffen- und Munitions-fabriken and began to turn out weapons and precision instruments of high quality.

There is reason to suspect that Lord Rothschild may already have had in mind an eventual amalgamation with Vickers when he introduced Sigmund Loewe to the board of Maxim-Nordenfelt and thus brought him into contact with Albert Vickers. Loewe, an indefatigable worker and a stimulating companion, made a highly favourable impression on the Vickers family. The amalgamation duly took place in the year in which Vickers acquired the Naval Construction and Armaments Company and Armstrongs absorbed Whitworths. On top of the £425,000 paid for the Barrow shipyard Vickers had to find £1,350,000 for Maxim-Nordenfelt, but the double merger made them possessors not only of valuable plant at Sheffield and Barrow but also of two machine-gun factories and an ammunition factory in Kent and of ranges in

Kent and Cumberland suitable for the testing of all kinds of weapons from those of small-arms calibre to the heaviest naval guns.

To finance these transactions, Vickers issued debentures to the value of £1,250,000 and increased their issued capital to £2,500,000, more than half of it in preference shares or preferred stock. The company's name was changed to Vickers, Sons and Maxim Limited, but became Vickers Limited when Maxim left the board in 1911.

The outcome of the Vickers-Barrow-Maxim and Armstrong-Whitworth mergers was that by the beginning of the twentieth century there were two British armament firms comparable with Schneider et Compagnie in France and Friedrich Krupp AG in Germany. Among American firms, the Bethlehem Iron Company and Carnegie Phipps were keen competitors with Vickers and Armstrongs in the international market for naval armaments and armour, and E. I. Du Pont de Nemours and Company was a large producer of propellants and explosives. Apart from Skoda, other armament undertakings which became of some importance by the outbreak of the First World War included enterprises in Italy, Japan, Russia and Turkey, some of them established with the help of British capital and expert guidance.

After their acquisition of the Naval Construction and Armaments Company and Maxim-Nordenfelt, Vickers branched out on a considerable scale. Between 1897 and 1914 they invested about four million pounds in subsidiary or associated companies at home and abroad.

The oldest Vickers subsidiary was the Placencia de las Armas Company, acquired as part of the deal with Maxim-Nordenfelt. Although its factories were in Spain, this was not a Spanish but an English company, set up by Nordenfelt in 1887 to manufacture arms for the Spanish government. Vickers took over from Maxim-Nordenfelt the whole of the share capital. The company lost money fairly steadily until 1910 and had to be heavily subsidized by its new owners. It then began to receive large orders from La Sociedad Española de Construccion Naval, an undertaking established by the Spanish government, with help from Vickers and John Brown and Company, to give Spain a new navy.

In 1901 Vickers founded the Wolseley Tool and Motor Car Company, a wholly-owned subsidiary which for some years made substantial profits by manufacturing and selling motor-cars and which also manufactured military vehicles. In the following year they bought half the share capital of William Beardmore and Company, a highly-respected Glasgow shipbuilding firm which specialized in armour plate and was threatening to compete with Vickers as a manufacturer of naval ordnance. When the Whitehead Torpedo Company's factories at Weymouth and Fiume came unexpectedly on the market in 1906 they joined Armstrongs, at the urgent request of the Admiralty, in taking them over on terms which gave them and Armstrongs a controlling interest in the company and members of the Whitehead family a minority holding. In 1905 and 1906 they provided roughly a quarter of the issued capital of the Italian-controlled Vickers-Terni Società Italiana d'Artiglieria ed Armamente. At the same time they bound themselves to give the Italian company the full benefit of their knowledge and experience in return for a percentage of the profits.

In 1910, by arrangement with the Canadian government, Vickers began the construction at Montreal of a shipyard intended to provide Canada with a home-built navy. A company called Canadian Vickers Limited was formed to manage the enterprise. Extensive workshops and a totally enclosed berth were built, and in 1912 a floating dry dock was completed at Barrow, towed across the Atlantic and delivered to the site.

Between 1907 and 1914 Vickers also contributed capital and expert guidance to the establishment, reorganization or administration of substantial undertakings in Russia, Japan and Turkey.* Not only in connection with the Whitehead affair but also in Japan and Turkey they acted in partnership with Armstrongs, just as in Spain they had acted in partnership with John Brown and Company. This may seem surprising in view of the rivalry between the two firms, but such paradoxes were not uncommon. For many years Vickers and Armstrongs maintained a working arrangement by which, when tendering for foreign contracts, they agreed in certain cases to concert their bids on the understanding that, should either of them be successful, the other would receive a share of the profits. This was one of those arrangements which critics of the arms trade would certainly have regarded as calculated to increase the price of armaments to governments. It did

* See Chapter 8.

not, however, operate to the disadvantage of any government to which the parties owed allegiance, and it did not give them anything approaching a monopoly. Whether they concerted their bids or not, Vickers and Armstrongs still had to compete with the rest of the world.

Apart from such special understandings, in many instances arms manufacturers were linked before the outbreak of the First World War by arrangements between individual fiims for the exchange of technical information, the pooling of patents and the granting of licences, or by common interests such as those which led to the formation of the Harvey Syndicate, the Steel Manufacturers' Nickel Syndicate and the Union des Mines. Such arrangements were not, of course, by any means confined to the arms trade. They were—and still are—a common feature of specialized trades in which all the participants use processes based on a relatively small number of inventions and discoveries.

In 1902, for example, Vickers acquired from Friedrich Krupp AG the right to manufacture, and to sell at home and abroad, time fuses and time-and-percussion fuses covered by Krupp patents. These were used in shells made by Vickers, and the royalty was added to the price. This was a perfectly normal arrangement with no sinister connotations, but it afterwards gave rise to a good deal of ill-feeling because the British government, preoccupied with more important issues, waited until more than half-way through the First World War before suspending the agreement between Vickers and Krupp under the terms of the Trading with the Enemy Act. Uncertainty then arose as to whether suspension of the agreement meant that it had become inoperative on the outbreak of war. Eventually Vickers refunded about £300,000 charged to the government for royalties on fuses in shells already supplied and paid for, receiving in return an indemnity with respect to any liability to Krupp for the period in question, insofar as this might not be cancelled by the terms of the peace treaty. The outcome was that in 1921 Friedrich Krupp AG claimed £260,000 for fuses sold between 1914 and 1917. Their claim was referred to an Anglo-German Mixed Arbitral Tribunal set up for the purpose of deciding such issues. In 1926, after the tribunal had adjourned without reaching a decision, the matter was settled by an agreed payment of £40,000 made by Vickers to Krupp. But some members of the public were left with the impression that the British government had had to pay royalties to Krupp on all the shells fired by British troops at the German lines throughout the war.

Arrangements between British and German firms for the manufacture of automatic weapons under licence had a still longer and more complex history. By 1896 Maxim-Nordenfelt were making both the ·303-inch Maxim gun and a 37-millimetre version known at the time of the South African War and later as a pom-pom. In that year Deutsche Waffen acquired a licence to manufacture Maxim guns in Germany, but this was subject to an interest acquired by Krupp before Sigmund Loewe became Managing Director of Maxim-Nordenfelt and Ludwig Loewe set up Deutsche Waffen. In 1901, after Vickers had absorbed Maxim-Nordenfelt, a new arrangement was made by which Deutsche Waffen were granted, with Krupp's consent, an exclusive licence to manufacture in Germany automatic and semi-automatic weapons with a calibre of 37 millimetres or less covered by Vickers patents, and the right to sell them anywhere outside France, the French colonies, the British Empire and the United States. Vickers claimed no royalties after the First World War on guns sold during the period of hostilities, but lodged two small claims with respect to machine-guns and parts for machine-guns sold earlier. Deutsche Waffen countered by claiming £75,000 under a loosely-worded clause in the agreement which entitled them to a share of the profits on sales they had helped to promote. Vickers, alleging that there was independent evidence that Deutsche Waffen had sold many more guns before the war than they admitted to selling, then put in a much larger claim for royalties, adding that they were unable to say exactly how much they were owed because the returns made by Deutsche Waffen were, in their opinion, fraudulent. Eventually both sides withdrew their claims and Vickers paid Deutsche Waffen £6,000 'in full settlement of all accounts'.[4]

The submarine, like the machine-gun, was a weapon of war which owed its origin to a relatively small number of inventors. We have seen that by the end of the 1880s Nordenfelt succeeded in making it an article of commerce to the extent of selling one submarine to Greece and two to Turkey. The steam-driven submarines built for Nordenfelt at Barrow were substantial craft, a hundred feet and more in length and carrying a crew of six or nine. But a weakness of the Nordenfelt submarine was that, although it could be made to dive at the will of its commander when under power, it soon rose to the surface if the power was cut off. To remain submerged, it had to consume fuel by maintaining an appreciable forward speed.

J. D. Holland, an Irishman who planned from 1878 to use submarines built with funds contributed by Irish-Americans to blow up British warships in harbour and thus make Old Ireland free, designed a submarine which did not suffer from this weakness. In 1898 Isaac Rice, a New York lawyer who had made a fortune from electric storage batteries, became interested in a submarine completed by Holland in the previous year. Two years later he persuaded the United States Navy to buy it. In the meantime he formed a company called the Electric Boat Company to exploit Holland's invention. Holland assigned his patents to the company, but afterwards quarrelled with Rice. He ceased in 1904 to take an active part in the company's affairs.

Rice made up his mind to seek markets outside the United States. He had met Sigmund Loewe in New York in 1898 and had many friends and associates among New York bankers and brokers. Armed with an introduction to Lord Rothschild, he sailed for Europe in the summer of 1900. In the following October he concluded with Vickers, to whom he was recommended both by Rothschild and by Sigmund Loewe, an agreement which gave them a 25-year licence covering the whole of Europe and the right to grant sub-licences to other firms, or to governments or states, on terms to be agreed with the Electric Boat Company.

In 1901 the first of a series of 63-foot-long experimental submarines based on Holland's designs was launched at Barrow. Vickers delivered five of these small craft to the Admiralty by the beginning of 1903. They went on to complete by 1910 fifty-six larger and faster submarines of the A, B, and C classes, with maximum speeds on the surface of twelve knots or so, as compared with the eight knots of the experimental craft. Long before the last of the C-Class boats was built, a start was made with still larger and faster submarines of the D class. These were 160 feet long, could make about fourteen knots and had improved seagoing qualities, but were difficult to manoeuvre when submerged. The Admiralty ordered broader-beamed boats of the E class, but in 1911 gave Vickers two years' notice of the termination of their contract on the ground that deliveries were too slow. Vickers then enlarged their plant, and in 1913 and 1914 they received further orders. In the meantime they and a rival firm, Scotts' Shipbuilding and Engineering Company, were invited to submit designs for short-range and long-range submarines which would meet requirements laid down by a committee of naval officers. Eventually Vickers, as well as Scotts and Armstrongs, were also invited to tender for the construction of submarines to an Admiralty design. But all seventy-five of the submarines with which Britain entered the war in 1914 were built to designs developed by Vickers from the Holland boat, and all except a few built in the naval dockyard at Chatham came from Barrow.[5]

In the early stages, progress at Barrow far outstripped that made by the Electric Boat Company on the other side of the Atlantic. For some years the company received very few orders from the United States Navy, with the result that it was soon in low water. Vickers helped with loans, and also by buying enough shares from minority shareholders to ensure that Rice would not be outvoted at general meetings of the company as long as they continued to support him. Eventually the company did receive government orders which enabled it to repay the loans, but the royalties it received from Vickers on submarines built for the Admiralty remained an indispensable part of its revenue. These royalties were substantial enough to become something of a burden to the British—or at any rate to Vickers—at a time when other countries managed to build quite satisfactory submarines without paying tribute to the Electric Boat Company.

Unlike submarines, airships and aeroplanes were not regarded as instruments of war when they were first introduced. Man's first flight in a powered aircraft was made in 1852, when a steam-driven airship carried Henri Giffard safely from Paris to Trappes. The first successful flight by a large rigid airship followed, after nearly half a century, in 1900. Man-carrying powered aeroplanes designed by Clément Ader and Wilbur and Orville Wright made short flights in 1890 and 1903 respectively. But it was not until six or seven years before the outbreak of the First World War that armies and navies paid much attention to aircraft other than balloons. By that time the nucleus of an aircraft industry already existed in France and elsewhere. At least half-a-dozen British, French and Italian firms, some with considerable experience of balloon-making, were capable of producing small pressure airships suitable for tactical or maritime reconnaissance. An American firm, the Goodyear Tire and Rubber Company, was not far behind them. In Germany two firms, Zeppelin and Schütte-Lanz, were manufacturing large rigid airships by 1909, and pressure airships designed by August von Parseval and Nikolas Basenach were being made in that year at a private and a government factory respectively. As early as 1905, Gabriel and Charles Voisin set up an aeroplane factory at Billancourt, on the outskirts of Paris. Beginning with towed gliders, they soon turned to powered aircraft.

Progress thereafter was rapid. For a time the Voisins worked in association with Henry Farman, a British subject domiciled in France. Before long he and his brother Maurice Farman branched out as independent manufacturers. The Wright brothers, after offering their invention without success to the United States, British and French governments, granted licences to European firms. In 1908 they demonstrated their Wright A biplane on both sides of the Atlantic. It was far more manoeuvrable than any contemporary European biplane, but could be flown only by a pilot who had mastered the art of handling an aircraft designed, like a bicycle, to be inherently unstable.

Rival designers concluded that they had much to learn from the Wrights but could contribute something of their own. In 1909 both Henry Farman and the American Glenn Curtiss produced biplanes much easier to fly than the Wright A and a good deal more manoeuvrable than the Voisin and Voisin-Farman biplanes of past years. In addition, successful monoplanes were produced in France by Léon Levavasseur and Louis Blériot and in Austria by Igo Etrich and F. Wels. In the same year Hans Grade, the first German-born pilot of a heavier-than-air machine, flew two different aircraft designed by himself, and successful flights by aeroplanes built in England were made by J. T. C. Moore-Brabazon, A. V. Roe and the Texan S. F. Cody.

Governments confronted by the Agadir crisis with the threat of a war in which France, Russia and perhaps Britain would be matched against Germany and Austria had some reason, therefore, to suppose that private enterprise should be able to provide most of the aircraft they would need during hostilities thought unlikely to last more than a few months. So far as aeroplanes, aero-engines and small pressure airships were concerned, the French had a big start over all competitors. The Russians, although they claimed that a monoplane designed by A. F. Mozhaiski in the 1880s was the first man-carrying powered aeroplane to fly in any country, were a long way behind them.* But they were willing to pay foreign experts to show them how their productive capacity could be expanded, and the Russian designer Igor Sikorsky was ahead of most of his contemporaries in foreseeing a future for large biplanes as bombers or transport aircraft. The Germans had, for the time being, a monopoly of large rigid airships. They and the Austrians were latecomers so far as the production of heavier-than-air machines was concerned, but by 1914 Germany would be able to count not only on her own resources but also on those of the Dutch designer and aircraft manufacturer Anthony Fokker, a genius who built his first aeroplane in 1911 at the age of eighteen.

Britain was an enigma. The Liberal statesmen who sanctioned staff talks with the French in 1906 were unable to say five years later whether, or in what circumstances, British troops would join the French in repelling a German invasion of France. Nevertheless an expeditionary force formerly intended for despatch to India in the event of war with Russia was made ready for despatch to the European mainland. A plan was made to concentrate it at Maubeuge and deploy it on a front facing east in the Ardennes, although in the outcome it had to be deployed on a front facing north at Mons. The Royal Flying Corps was formed in 1912 to provide the necessary air support. An Army Aircraft Factory which had taken the place of the old Army Balloon Factory at Farnborough was then renamed the Royal Aircraft Factory. The government's intention was that it should not compete with private firms by manufacturing aeroplanes or aero-engines in commercial quantities, but should co-operate with them by establishing qualitative standards and by testing aircraft and engines acquired by or offered to the state. It was not, however, precluded from building experimental aircraft. Under the direction of a technical staff which included the future aircraft manufacturer Geoffrey de Havilland, the Royal Aircraft Factory produced

* Mozhaiski's aircraft, powered by a British-made steam engine, travelled about thirty yards through the air after reaching the bottom of an elevated ramp. Whether it could have sustained itself in level flight is debatable.

prototype versions of a wide range of aircraft afterwards manufactured by commercial firms on government account.

The government recognized that there were some forms of aircraft construction which no commercial firm was likely to undertake on its own initiative and which only a large firm could tackle. When the authorities decided in 1909 that a large rigid airship should be built for the Admiralty, they turned to Vickers as the only British firm with experience of the construction both of surface warships and of submarines. Vickers, already asked by the Admiralty in the previous year to quote a price for an airship of Zeppelin type, agreed to build the airship afterwards called the *Mayfly*. Captain R. H. S. Bacon, Director of Naval Ordnance and one of the chief sponsors of the project, was to have supervised the construction of the ship, but he was posted away from the Admiralty as an indirect consequence of a high-level dispute about strategy. Captain Murray Sueter then accepted the post of Captain Inspector of Airships on the express understanding that he was to have no responsibility for design.

This did not mean that Vickers were given a free hand. They had already designed the main girders for an airship of Zeppelin type when they learned that the Admiralty wanted an airship considerably larger than any Zeppelin yet built. So many departures from the original concept had to be made to meet the Admiralty's requirements that H. B. Pratt, a mathematician employed by Vickers, predicted that the structure would collapse unless further and more radical changes were made.

The *Mayfly* was hauled from her shed at Barrow for buoyancy and handling tests on 22 May 1911. She behaved well, but was pronounced too heavy by three tons. Modifications were then made to improve her lift.

When the *Mayfly* made her next appearance in September, Pratt's forecast proved all too correct. Within a few minutes of her emergence she broke her back and became a useless wreck.

Vickers, having spent £50,000 on the construction and equipment of a shed big enough to hold a large rigid airship, recognized even before the *Mayfly* was wrecked that they would need further orders for aircraft of one kind or another if they were not to lose by their incursion into a new department of the arms trade. Murray Sueter advised them to turn to aeroplanes rather than count on the Admiralty to buy more airships. He mentioned two

French aircraft, the Sommer biplane and the REP monoplane, which might prove suitable.

The designer of the Sommer biplane turned out to be unwilling to do business with a British firm. Robert Esnault-Pelterie, designer of the REP, was more amenable. As the outcome of conversations between Esnault-Pelterie and the company's technical adviser, Zaharoff negotiated an agreement which gave Vickers an exclusive licence to manufacture and sell REP aeroplanes and aero-engines in the British Empire.

The first monoplane built by Vickers to Esnault-Pelterie's design was completed at Erith in the summer of 1911. Two years later the company exhibited at Olympia a two-seater pusher biplane, the FB.1, designed for it by G. H. Challenger. This had a machine-gun mounted in the nose. Altogether, the company completed twenty-six aircraft by August 1914.[6]

Armstrongs, too, entered the aircraft industry before the outbreak of war. In 1913 they set up an aircraft department to manufacture for the War Office biplanes designed at the Royal Aircraft Factory. Later they built aircraft to their own designs, but the first of these did not appear before 1915.

The contract with Esnault-Pelterie was one of many transactions conducted by Zaharoff on the company's behalf. Besides selling Vickers armour abroad from the early 1890s, he had continued until the amalgamation of 1897 to act for Maxim-Nordenfelt. The amalgamation made him an agent for Vickers, Sons and Maxim Limited on terms which entitled him to nine-tenths of one per cent of the company's profits. This arrangement ran from 1896, although the amalgamation was not completed until the following year.

Conflicting accounts of the circumstances in which Zaharoff came to be acquainted with members of the Vickers family will be discussed in the next chapter. Here it is enough to note that before 1896 he was employed by the family firm to sell only armour, and at all times was employed solely in connection with foreign business. In 1927 Vickers presented him with a silver-and-gilt cup 'on completion of 50 years of service'. This might be thought to establish 1877 as the date when he joined the firm; but Vickers were not then arms manufacturers and would have had no use for the services of an arms salesman. Presumably the presentation was intended to mark the fiftieth anniversary of the date when Zaharoff entered the arms industry by joining Nordenfelt's company, afterwards absorbed by Vickers.

A few months after the date in 1898 when the terms of Zaharoff's employment by Vickers, Sons and Maxim Limited were confirmed by a formal entry in the company's records, the directors came to a new arrangement by which he received a remuneration equivalent to ten per cent of that of the Managing Directors and a contribution of £1,000 a year towards the upkeep of his house in Paris. In 1901 he was empowered to draw on the company for his expenses and the cost of maintaining an intelligence department at a rate equivalent to one per cent of the gross selling value of the output of war materials from the company's factories in England and Spain. Four years later his remuneration as 'foreign adviser' was fixed 'until further alteration' at a sum equivalent to the bonus to which the Junior Managing Director was entitled under the company's Articles of Association.

The practical effect of these arrangements was that Zaharoff drew from Vickers, Sons and Maxim Limited or Vickers Limited, between 1902 and 1913, about £640,000 by way of commission, remuneration and allowances for disbursements.[7] Figures for earlier years are not complete, but a fair estimate is that altogether he must have drawn, between 1897 and the outbreak of the First World War, about three-quarters of a million pounds. In 1902 and 1903 the directors sanctioned payments amounting (to the nearest pound) to £33,667 and £35,349 respectively as 'special commissions'; in 1913 they sanctioned payments of £49,193 as commission, £12,894 for expenses and £2,493 in connection with Spanish business, making a total of £64,530 for work done and expenses incurred in 1912. In one year Zaharoff received as much as £85,771; in two other years payments made to him exceeded £80,000. Furthermore, by 1913 the company was paying him tax-free interest on a loan of £115,025.

These would be large sums, even today. By Edwardian standards they were immense. Who was this Mr Zaharoff who handled so much money, lived in the most fashionable part of Paris, owned a country house in France and was believed by some foreigners to be no mere employee of Vickers but the company's 'principal shareholder'?

FIVE

The Man of Mystery: Zaharoff

For a good many years before his death in 1936, Basil Zaharoff was known in England by the style and title of Sir Basil Zaharoff. Aware of his long association with the British arms industry, probably most Englishmen who read in their newspapers of his comings and goings assumed him to be a naturalized British subject. Since the name transcribed in Western European languages as Zakharov, Zacharoff or Zaharoff is not uncommon in Russia and is said to be derived from the Hebrew Zohar, it is not surprising that some took it for granted that his forebears were Russian Jews.

Zaharoff was in fact a naturalized Frenchman, long domiciled in France. So far as his nationality of origin was concerned, he claimed to be a Greek born in Turkey of Christian parents. Basil was an anglicized version of his second name, Vasiliou. His English friends, to whom he sometimes wrote letters signed Zed Zed or Zedzed, knew that his first name was Zachary or Zacharie. His knighthood was genuine, he had accounts with the Bank of England and at least one other bank in London, and he belonged to a well-known London club. But whether, in view of his French citizenship, he ought properly to have been styled Sir Basil Zaharoff or M. Basile Zaharoff remains a nice point for students of protocol.

An aura of mystery surrounds Zaharoff's birth. In 1908, on the eve of his appointment as a Commander of the Legion of Honour, he exhibited to a French court, in lieu of a birth certificate, a document which purported to have been drawn up on 21 December 1892, at Mouchliou, or Mughla, in Anatolia and was signed by three witnesses.[1] It was countersigned by the Archimandrite Makharios as witness to the authenticity of the three signatures, and by the Ecumenical Patriarch Neophytos as witness that the Archimandrite's signature was genuine. Whether it was drawn up to support an application for French citizenship made in or about 1892 is not clear, but presumably it was intended for some such purpose. According to a certified French translation from the original Greek, the witnesses testified that Zacharie Vasiliou Zacharoff was born on 6 October 1849, at Mughla, was baptized two days later by 'the priest Daniel', and was the legitimate son of Vasiliou and Hélène (or Helena) Zacharoff. The register in which the birth and baptism were recorded was said to have perished in a fire which destroyed the priest Daniel's church.

The same date and place of birth were mentioned many years later in connection with the proving of Zaharoff's will.[2] On the other hand, Zaharoff appears to have given his age in 1873 as twenty-two—which would place his birth in 1850 or 1851—and to have made in 1921 a statement which implied that his birthplace was Constantinople. There is also the testimony of a friend of the Vickers family who said about 1962 that she knew for a fact that Zaharoff was born in Manchester.[3] She mentioned a photograph of Zaharoff, taken at a very early age, which bore the imprint of a Manchester photographer. She admitted, however, that two younger sisters, who also appeared in the photograph, were understood to have been born not in England but in Turkey.

Unfortunately, none of this evidence is conclusive. The authenticity of the document allegedly drawn up in 1892 has been questioned on the ground that the witnesses, unless prompted, were not very likely to remember the precise dates of birth and baptism of a child born more than forty years earlier. The fact that the same date and place of birth were mentioned when Zaharoff's will was proved is not significant, for the two distinguished Frenchmen who testified to the identity of the deceased can have had no independent knowledge of the circumstances of his birth. Zaharoff's age may have been correctly stated in 1873, but a mistake or an intentional misstatement cannot be ruled out. The lady who said she knew for a fact that Zaharoff was born in Manchester cannot really have known it for a fact, because she was not alive at the time and did not claim to have seen any documentary evidence more convincing than a photograph taken in early childhood. Finally, it is not absolutely certain that Zaharoff meant to be understood in 1921 as claiming that he was *born* in Constantinople. The context shows that the point he was concerned to make was that he was brought up there.

However, if any weight at all is to be given to hearsay evidence it seems clear that Zaharoff did spend part, at least, of his boyhood in Constantinople.[4] According to statements made many years later by people who claimed to have known the family, the Zaharoffs were poor. They lived not in the prosperous quarter mentioned by Zaharoff in 1921 but in one notorious for its poverty. A benefactor whose name has been rendered, not very convincingly, as Iphestidi, is said to have paid for the boy's education at 'the English school'. This assertion has been challenged on the ground that there were no English schools in Constantinople at the relevant time except mission schools, which were free. But there were plenty of English schools in England. In the light of subsequent events, it seems not inconceivable that the mysterious Iphestidi, who is thought to have been a kinsman of Zaharoff's mother, may have arranged for him to spend a year or two in England with a view to his learning enough about the country to fit him for eventual employment as London agent, correspondent or manager of a family business. In his maturity Zaharoff wrote good if rather flowery English, though he sometimes put his name to letters written in execrable English by subordinates.[5]

Nothing like a clear picture of Zaharoff's early years emerges from the many thousands of words written about him by unauthorized biographers. He is said to have begun his commercial career as a guide and money-changer, but all such statements—including the assertion that he was first employed as doorkeeper at a brothel—must be viewed with caution as emanating from hostile witnesses. A Birmingham man named Haim Manelewitsch Sahar, who afterwards called himself Hyman Barnett Zaharoff, claimed many years later to be the offspring of a marriage contracted by Zaharoff in 1867 in accordance with the rites and ceremonies of the Jewish community of the Ukraine.[6] He added that the marriage was annulled in 1875 on the ground that it was not preceded by a formal betrothal. Zaharoff would have had to be a very youthful bridegroom to make the story true, but early marriages are said to have been not uncommon in Tsarist Russia as a means of escaping military service. However, by the time the claim was made Zaharoff was already a rich man and had formed an attachment to a married woman, the Duquesa de Marchenas, whom he hoped to make his wife as soon as her supposedly sickly husband died. Even if we accept a version of the story which holds that he silenced Sahar by setting him up in business in London and making provision for his daughters, it does not necessarily follow that Sahar's claim was valid. Zaharoff could have known that it was false and still have felt that surrender to blackmail was preferable to a resounding scandal.

Zaharoff afterwards denied that he was in or had ever visited the Ukraine at the time when he was alleged to have married Sahar's mother. At all events, it seems clear that by 1873 he was in England. In that year a young man described as Zacharia Basileos or Basilius Zacharoff, a Greek aged twenty-two, was charged in London with an offence new to the statute book. He was accused of pledging, as agent, boxes of gum and sacks of gall entrusted to him by his principal, a Constantinople merchant whose name was given as Manuel Hiphentides. He appeared before the Recorder, pleaded not guilty, but changed his plea to guilty when he was told that a sum of money in his possession could not be used to bring a witness from Constantinople. The Recorder, evidently feeling that he had not got to the bottom of the matter, postponed sentence, and at a subsequent hearing the defendant was bound over and released on the understanding that any loss incurred by the complainant would be made good.[7]

Zaharoff appears not to have denied in later years that he was the defendant in this case and to have offered a number of explanations of the affair. The gist of them was that he was not an agent but a principal. Had he persisted in his plea of not guilty, he would have testified that he did not pledge goods entrusted to him but merely helped himself to money due to him under a partnership agreement. According to one version of the story, after the trial had begun he found his copy of the agreement in the pocket of an overcoat he had not worn for some time, but hesitated to produce it because his accuser was his uncle and he did not wish to expose a member of his family as a perjurer.

Bearing in mind the nationality of the parties and the peculiarities of Greek family life, the reader may feel that this explanation is not altogether unconvincing. One does not have to accept the story in its entirety to believe that there may have been at least a grain of truth in it. If Manuel Hiphentides really was Zaharoff's uncle—and especially if he was the 'Iphestidi' said to have paid for Zaharoff's education—the charge brought against a nephew who had the temerity to help himself to money without express permission becomes understandable as the outcome of a family quarrel whose complexities no English court could have been expected to unravel.

Zaharoff appears to have spent the next few years in Greece and to have been caught up in the undercurrents of Greek politics. He is said to have been fond in his later years of telling far-fetched stories of his adventures as a kind of courier or liaison officer between aspirants to power and their clandestine supporters. These may not have been true, but there is no doubt that he acquired at some stage an exceptional knowledge of some of the more recondite aspects of Near Eastern politics.

According to a persistent legend in the Vickers family, his first meeting with a member of the family occurred at a time which must, on internal evidence, be assigned to this period. During a visit to Athens, Tom Vickers was sitting on the terrace of his hotel when he saw at a neighbouring table 'a very young man' who seemed nervous and embarrassed. Tom Vickers spoke to him, bought him a drink, and invited him to call should he ever find himself in England. The 'very young man' is supposed to have been Zaharoff.[8]

One version of the story asserts that Zaharoff followed up the invitation and was received not by Tom but by Albert Vickers, Tom's junior by three years. He asked to be given some kind of agency. When told, presumably at a subsequent interview, that the board had decided to grant his request, he fell on his knees, exclaiming: 'Mr Vickers, I am starving. This has saved my life!'[9]

Unfortunately this picturesque anecdote cannot easily be fitted into the framework of ascertained fact. We have seen that in 1927 Vickers Limited regarded Zaharoff as having completed fifty years' service. But that, presumably, was because the company had absorbed the Nordenfelt interests and the directors therefore considered themselves to be, in a sense, the heirs and representatives of Nordenfelt. So far as can be ascertained, there is nothing in the company's records to suggest that Zaharoff worked directly for Vickers, Sons and Company before the early 1890s, when he was employed to sell the company's armour in the course of his travels on behalf of Maxim-Nordenfelt. By that time he had been earning a good income as an arms salesman for many years and was an acknowledged authority on foreign markets. It is true that Maxim-Nordenfelt struck a bad patch in 1894 and gave him six months' notice, but he cannot conceivably have been starving. Moreover, at any time after the Maxim-Nordenfelt amalgamation of 1888 he could easily have gained access to Albert Vickers without having to invoke the memory of a supposed meeting with Tom Vickers many years earlier. His name repeatedly cropped up in the minute books of Maxim-Nordenfelt as that of a trusted agent, and Albert Vickers must have been familiar with it.

What, then, was the job which rescued Zaharoff from penury and was given to him by a board of directors whose spokesman was Albert Vickers? It can scarcely have been the agency for Nordenfelt which he acquired in the 1870s. Nordenfelt's English company was not formed until 1886, and the Vickers family did not become associated with it until the amalgamation of 1888 united it with the Maxim Gun Company. Zaharoff was one of the foreign agents of Nordenfelt whose services the Barrow Shipbuilding Company was authorized in 1887 to call upon, but Albert Vickers was not then in a position to speak for either Nordenfelt or the shipbuilding company. He may have ratified Zaharoff's appointment as an agent for Maxim-Nordenfelt after the amalgamation, but that would not have been much more than a formality. It is hard to believe that Zaharoff went down on his knees to thank a director of the new company for re-engaging him in a position he had held under the old dispensation.

There is thus an unresolved conflict between the conviction of members of the Vickers family that Zaharoff had a fruitful meeting with Tom Vickers while he was still a very young man, and the absence of any record of his having worked directly for Vickers, Sons and Company before he was forty or more. It is, of course, quite possible that he did meet Tom Vickers in Athens in the 1870s and spoke of the meeting when, many years later, he found himself working for the family firm. Alternatively, the meeting may never have occurred but both men may have persuaded themselves that it did; or Zaharoff may have allowed the assumption that he was the youth Tom Vickers had met to pass unchallenged in order not to spoil a good story, or perhaps to ingratiate himself with the family. But none of these hypotheses would account for the anecdote which depicts Zaharoff as going down on his knees to thank Tom's brother for employing him.

At any rate it is not disputed that Zaharoff became Nordenfelt's agent or correspondent in the Near East in the 1870s, or that he worked for Maxim-Nordenfelt from 1888, was appointed Maxim-Nordenfelt's 'foreign adviser' in 1890 and for some years was also an agent for the armour introduced by Vickers, Sons and Company about that time. We have seen that, after the absorption by Vickers of Maxim-Nordenfelt and the Naval Construction and Armaments Company, he drew very large sums from Vickers and by 1913 was receiving interest on a loan of £115,025. The odd £15,025 was the price of some mining shares sold to the company, but how Vickers came to owe their foreign adviser the remaining hundred thousand pounds the record does not show and no one now remembers.

He was not, however, either then or later, the company's 'principal shareholder'. He was never the registered holder of more than a few thousand shares, and there is no reason to suppose that he ever held any large number of shares through nominees. Nor was he ever a director of Vickers, Sons and Company, Vickers, Sons and Maxim or Vickers Limited.

The fact remains that, besides being one of three representatives of Vickers on the board of Vickers-Terni, he was for many years the trusted adviser and close confidant of the men who directed the affairs of Britain's largest armament firm. He was consulted or kept informed about matters of importance to the company not only by the directors but also by the expert they called in when a drastic reorganization of the company's finances became necessary between the First and Second World Wars. He took a keen interest in many aspects of the company's business, even concerning himself with such domestic issues as staff catering and the relative merits of canteens and luncheon vouchers. He was on terms of intimate friendship with members of the Vickers family.

Clearly, the directors of Vickers did not view Zaharoff in the same light as did critics of the arms trade who saw in him only an unscrupulous, cynical exploiter of human weakness. Was this because they, too, were unscrupulous, cynical exploiters of human weakness? Perhaps. Even so, it is improbable that Zaharoff would have lasted long as foreign adviser to a firm with worldwide interests had he been no more than the flamboyant adventurer depicted by hostile biographers. Tom and Albert Vickers were hard-headed Yorkshiremen, born and bred in the briar-patch of the Sheffield steel industry. They were not guileless provincials, likely to be easily deceived by a cunning rogue. Tom, it is true, had few interests outside the business apart from chess, his duties as a local magnate and the volunteer regiment he commanded. Albert, who succeeded his brother as Chairman in 1909, was a man of the world who talked of great affairs 'in an easy, confident, off-handed way as matters, to him, of everyday occurrence'.[10] He liked good food, good wine and good company. He enjoyed shooting over the best grouse moors. Douglas Vickers, Chairman from 1918 until 1926, was a third-generation steel magnate with wide interests. He had a taste for the company of clever men and said he had been bored at Eton because he found the work too easy.

Moreover, there were other men close to the top besides these three. Sigmund Loewe, a director from 1897 until he was killed in a motor-car accident in 1903, was a cultivated German Jew, educated at a Roman Catholic monastery and a friend of the Rothschilds and their circle. Trevor Dawson, who joined the firm in 1896, was a former naval officer and gunnery expert remarkable for his good looks, air of distinction and trenchant wit. Vincent Caillard, a director from 1898, had held a commission in the Royal Engineers, had filled a number of posts in the Near East which placed him on the borderline between diplomacy and military intelligence, and was an expert on Turkish affairs. Francis Barker, who represented Vickers at conversations with the Board of Trade soon after the outbreak of the First World War, came of a banking family in Constantinople, where he was born. Educated in England, he had been confidential secretary to the Director-General of the British-controlled Ottoman Bank. Both he and Caillard knew Constantinople extremely well. They were well acquainted with the pecularities of Near Eastern commerce, finance and politics. Yet neither of them, it would seem, knew of any reason why Zaharoff should not act as their adviser and transact important business on the company's behalf.

This does not mean, of course, that they and their colleagues expected their go-between with foreign governments to abide by the standards they observed in their dealings with the British government. The company's prosperity between 1897 and 1914 depended to a considerable extent on its receiving orders for arms from parts of the world where large-scale business could seldom be transacted without bribery. Probably the directors did not know at all precisely how much of the three-quarter million pounds or so paid to Zaharoff during those years found its way into the pockets of corrupt ministers or officials. But it must have been within their knowledge that sums which figured in the accounts as expense allowances or special commissions included bribes distributed by Zaharoff in Russia, the Balkan states and presumably elsewhere. They could not fail to be aware that the payment of secret commissions to public servants in countries notoriously corrupt was common practice and that the only alternative to acquiescence in such abuses would have been to leave competitors with a clear field.

Men who made large incomes by the sale of arms were popularly supposed after the First World War to have viewed with satisfaction a state of affairs which was bound to increase the demand for their wares. Zaharoff was the most successful arms salesman of his day, but he was employed by Vickers solely in connection with foreign business. Unscrupulous though he may have been, it seems unlikely that he welcomed a war which, even if it lasted only a few months, might make the transaction of foreign business extremely difficult. The outbreak of hostilities did bring his employers orders from France and Russia, but priority had soon to be given to orders from the British government, on which he received no commission. At the beginning of 1916 he gave the company the notice required to terminate the arrangement by which he received an allowance based on output to cover his disbursements.[11] But wartime restrictions did not, as things turned out, prevent him from continuing to earn substantial sums. At the height of the war he obtained from the Russian government two large orders for machine-guns, tripods and spare parts. Vickers, unable to supply the goods because the old Maxim factory at Crayford had gone over to aircraft production and the former Nordenfelt factory at Erith was working at full stretch on orders for machine-guns from the British government, agreed that the Union Development Company of New York should do so and should pay Zaharoff a commission of two-and-a-half per cent on a contract price of more than twenty-seven million dollars.

The American company was unable to meet its delivery dates, and the contract was cancelled after roughly a third of the goods had been delivered. Zaharoff claimed the balance of his commission, arguing that the cancellation was no fault of his and that he had worked hard to prevent it. Vickers agreed with him. They paid him £95,000 and asked the Union Development Company, in a forceful letter, to reimburse them.[12]

Zaharoff's offer to forego his expense allowance received no formal answer for two years. At the end of that time Albert Vickers wrote to say that the directors looked back with great pleasure on the intimate and friendly relations that had existed between them and Zaharoff in the past and hoped that such relations would continue for many years to come. They suggested that, pending discussion, he should continue to draw on the company for his expenses at the average rate of the past two years.

Zaharoff was a fluent linguist, at home in half-a-dozen countries. He was an experienced collator and interpreter of industrial intelligence, operating in a field where industry overlapped with politics and high finance. An able and adroit negotiator, he understood the art of showing deference to an interlocutor without sacrificing his own or his employer's interests. He possessed not only exceptional knowledge of Near Eastern politics, but also the means of keeping it up to date. French journalists believed him to be a key figure in the British intelligence service. In England he was suspected of being the chief instigator and inspirer of Lloyd George's pro-Greek and pro-Venizelist attitude to Near Eastern problems. Without admitting that he was either, one can well believe that he did not owe his knighthood solely to his services to the arms industry.

Whether he played as important a part as he liked people to think in attempts to detach Turkey from the Central Powers during the second half of the war is another matter. Towards the end of 1917 Lloyd George, in consultation with the Foreign Office, sent General Smuts and Philip Kerr (afterwards Lord Lothian) to Switzerland to meet Count Mensdorff, a former Austrian ambassador in London and an unofficial emissary of the Austrian government. They were told that they should also make contact with Dr Parodi, head of an Egyptian mission in Geneva and understood to be an agent or spokesman of the Turkish government. On 18 December, while Smuts was closeted with Mensdorff, Kerr had a long talk with Parodi. The substance of what Parodi had to say was that the British could not hope to do anything in Turkey without the support of the ultra-nationalist Committee of Union and Progress, which continued in wartime to play the dominant role it had assumed when it engineered the Young Turk uprising in 1908. Enver Pasha, leader of the Germanophile section of the Committee, still believed that Germany could win the war. But there were also members of the committee who longed to break with Germany and would willingly 'lean on England' if they were satisfied that the Allies meant to grant Turkey moderate peace terms. Parodi strongly advised the British government to make its intentions clear not merely to them but to all members of the Committee, including Enver Pasha and his followers.

About three weeks later, Lloyd George made a speech in which he said that his government did not aim at depriving Turkey of Constantinople or 'the homelands of the Turkish race'. Published accounts of his attempts at peacemaking in the Near East do not clearly show what other steps he took to reassure the Committee of Union and Progress. So far as is known, Smuts and Kerr did not see Parodi when they returned to Switzerland in March to meet an emissary of the Austrian Foreign Ministry; but the emissary had visited Parodi a few days after Lloyd George's speech was made. Zaharoff claimed afterwards to have met Enver Pasha on a number of occasions in 1917 and 1918 and to have sounded him on behalf of the British government about the chances of concluding a negotiated peace. Lloyd George denied authorizing him to make such proposals. But that, of course, he was bound to do, and he would have taken care to put himself in a position to do it. The question is not whether Zaharoff was briefed by Lloyd George in person, but whether he received a brief from Downing Street through some such intermediary as Vincent Caillard or Francis Barker. If he did try his hand at mediation, with or without the authority of the British government, his efforts were not successful. The Turks did not ask for an armistice until their troops had been routed in Palestine and Syria and were on the verge of collapse in Mesopotamia.

Zaharoff continued after the war to draw large sums from Vickers. His remuneration as foreign adviser was fixed in 1919 at

£5,000 a year (increased in 1920 to £6,000 a year) and a bonus of three-quarters of one per cent of the company's profits. In 1922 the directors formally sanctioned payments totalling £761,250 made to him during the past four to five years to cover transactions since the beginning of 1915. These included repayment of large dollar advances made to the company in 1917.

In 1923 Zaharoff borrowed £150,000 from the company, apparently because currency restrictions prevented him from finding from his own resources the whole of a large sum in sterling needed for a particular purpose. He is said to have bought somewhere about that time a big block of shares—according to some accounts a controlling interest—in the company which operated the gambling concession at Monte Carlo and was understood to own the Hôtel and Café de Paris. Whether the loan was connected with that transaction is not clear. He repaid £50,000 when the loan matured in the following July, and was granted two six-monthly renewals of the balance.

From the beginning of 1927 Zaharoff was no longer paid an expense allowance based on the company's output of war materials, but received a flat sum of £10,000 a year. In addition to this allowance and his remuneration as foreign adviser, he received substantial commissions on orders placed by the Sociedad Española de Construccion Naval or the Spanish government with the Placencia de las Armas Company or with Vickers or its associated companies at home. For many years his receipts from Spanish business included a commission on every pound of powder bought for the Spanish Navy from Nobel's Explosives Company. This was paid by Imperial Chemical Industries Limited after the firm became a subsidiary of that company.

Long before the war Zaharoff had met in romantic circumstances, while travelling by train between Paris and Madrid, the newly-married wife of Don Francisco Maria Isabel de Borbon y Borbon, Duque de Marchenas. Her name before her marriage to Don Francisco was Maria del Pilar Antonia Angela Patronina Fermina Simona de Muguiro y Beruete, Duquesa de Villafranca de los Caballeros. The marriage was loveless and she was desperately unhappy. Zaharoff rendered her some small service. The meeting was the prelude to a lasting association. The Duke was a sick man and not expected to last long, but he confounded the prophets by maintaining a precarious hold on life for many years. It was not until 1924, when Zaharoff was in his seventies, that he was at last able to marry the woman whose faithful courtier he had been for almost as long as he could remember. He afterwards formally adopted her two daughters.

For many years Zaharoff divided his time, when he was not travelling, between the Avenue Hoche—where he occupied two houses in succession—and his country house in the department of Seine-et-Oise, the Château de Balincourt. The Château de Balincourt continued to be his home until his death, but he spent part of almost every year at Monte Carlo. People who saw him there remarked on his habit, when exposed to the public gaze, of casting anxious glances about him, as if he expected to meet an assassin or a press photographer at every corner. (Members of the Vickers family, it will be remembered, described the 'very young man' Tom spoke to in Athens as looking nervous and embarrassed.) This idiosyncracy was, perhaps, understandable in a man much exposed to criticism and even blackmail, whose background was obscure and who spent much of his life in fear of a scandal that might make further meetings with the Duquesa de Marchenas impossible.

Some of Zaharoff's contemporaries described him as an inveterate philanderer, others as so wholeheartedly devoted to the Duquesa that he had eyes for no other woman. Possibly his reputation as a ladykiller arose from attentions he thought it expedient to pay to the wives and daughters of customers. Again, friends have described him as a philanthropist who paid anonymously for the education of young artists; detractors as so mean that he would sit for an hour at the Café de Paris at Monte Carlo, order nothing but a bottle of mineral water and leave without tipping the waiter. Both could be right. Moreover, as a frequenter of the Café—perhaps even part-owner of it—Zaharoff may have had an understanding with the staff which relieved him of the necessity of ordering drinks he did not want and bestowing a tip at every visit.

One thing which does seem quite clear about Zaharoff is that the business to which he devoted his working hours was far more to him than a source of livelihood. To persuade a customer to give him a large order, claim his commission, make sure that it was paid punctually and that his sub-agents received their due—these employments brought him a satisfaction derived by other men from indulgence in their hobbies. If he seemed to have few recreations, perhaps one reason was that his daily round was so congenial that he did not need them. He enjoyed travel, negotiation, the striking of a bargain, as other men enjoy golf or sailing. So why play golf or sail? At one time he was persuaded by his English friends to take up shooting, and bought a pair of Purdey guns. But the great arms salesman found the recoil troublesome and soon gave them away.

On the eve of his death, Zaharoff was widely regarded as a multi-millionaire, the owner of property and investments worth, perhaps, as much as twenty million pounds. But this turned out to be a wild exaggeration. A man of mystery to the end, he left to be divided between his adopted daughters not the immense fortune he was thought to have accumulated by many years of artful salesmanship and shrewd investment, but approximately a million pounds. Commissions on Spanish business continued for some years after his death to be paid to his estate.

SIX

Armageddon

Few people believed on the outbreak of the First World War that hostilities between the leading European powers could or would last more than a few months. Moltke's plan for a two-front war against France and Russia was based on the assumption that, by marching his right through Belgium, he would be able to encircle the main body of the French Army by the thirty-ninth day of mobilization and would be ready by the following day to start passing divisions to the Eastern Front.[1] The Russians were known to be trying to modernize their arms industry, were not expected to be ready for war before 1916, and were thought unlikely to hold out for any great length of time against the combined might of Germany and Austria.

Moltke was given to periodical fits of gloom in which he feared the worst. He was warned that Belgian resistance, especially if stiffened by British intervention and active help from the French, might delay or even halt the advance of his armies of the right. The forts commanding the crossings of the Meuse at Liège and Namur, if held by determined men, would be formidable obstacles. But Moltke hoped, in his more sanguine moments, that diplomatic pressure and a show of force would induce the Belgians to grant unrestricted passage to his troops. Should they refuse to do so and an attempt to seize the Liège forts by a *coup de main* prove unsuccessful, he would fall back on Krupp's giant howitzers and the outsize mortars borrowed from the Austrians.

The French were equally hopeful. They had acquired, by devious means, an alleged copy of the Schlieffen plan which purported to bear Moltke's annotations. In any case there could be little doubt that the Germans would aim at marching through Belgium if it suited them to do so. But the Belgian forts were thought unlikely to fall except after a prolonged siege, and an unsound estimate of the fighting value of Moltke's reserve formations led the French to believe that he would be unable to extend his right even as far north as Namur, let alone Liège, without so weakening the rest of his line that they would be able to break through his centre. A German officer departing for the front in August 1914 spoke of breakfasting on the anniversary of Sedan at the Café de la Paix in Paris.[2] The French, not to be outdone, looked forward to celebrating Christmas in Berlin.

There were, of course, many people who did not believe either that the Germans would reach Paris by September or that the French would be in Berlin by December. But it does not necessarily follow that they expected a long war. Economists believed that, if neither side gained a swift decision, the practical disadvantages of a state of hostilities between the leading European powers would soon become so obvious that no statesman in his right mind would wish to prolong the struggle. One side or the other would then propose a negotiated peace.

An interesting point about this prediction is that it very nearly proved correct. The Belgians did refuse free passage to Moltke's armies, the *coup de main* was unsuccessful, but the Liège forts soon succumbed to plunging fire from the howitzers. Even so, the last of them did not fall until six days after the German First Army, on the extreme right, could have begun its advance if no resistance had been met. The First and Second Armies were afterwards further delayed, albeit briefly, by encounters with the British Expeditionary Force at Mons and Le Cateau and the French Fifth Army on the Sambre and the Oise. A premature turning movement then exposed the First Army to a flank attack from Paris. The result was that the thirty-ninth day of mobilization, which ought to have seen the First and Second Armies closing in on a defeated enemy, found the Second Army in a situation described by Moltke's emissary as 'serious but not hopeless' and the First Army in a still worse plight. On the following day both armies retreated from the Marne 'in wild haste' and Moltke wrote to his wife to say that the war was lost. Germany, he thought, would have to make peace and offer to pay for the damage she had done.

Unfortunately for all concerned, his superiors took a different view. Refusing to admit defeat, they dismissed Moltke—although his supersession was not made public until later—and condemned

Europe to four years of futile carnage by decreeing that their troops should stand fast on the Aisne.

By that time the French had shed some of the illusions with which they entered the war. Disastrous attempts to take the offensive in Lorraine and the Ardennes had exposed the falsity of their belief that troops charging with the bayonet, and supported only by light artillery, would be able to storm entrenched positions before the defenders had time to bring aimed fire to bear against them. Before the war the leaders of the French Army had given only grudging orders for weapons heavier than the 75-millimetre field gun to which they pinned their faith, because they believed that such weapons would impair the mobility of their formations. They now saw that, to reduce the enemy's positions, their troops needed medium and heavy guns as well as field pieces, far larger quantities of ammunition of all calibres than were provided for in peacetime estimates, above all large numbers of shells filled with high explosive. They also needed machine-guns and mortars to attack the enemy's trenches at close quarters and to defend their own positions; better entrenching tools than those hitherto thought adequate; new uniforms to replace the scarlet trousers and long blue tunics which made the French infantryman so appallingly conspicuous on the battle-field. All these things, and many others, had to be produced in quantities and at a speed which called for a supreme effort.

The France of 1914 was largely a land of peasants, artisans and craftsmen. France had become since the Napoleonic Wars a great industrial power, but she was still not as highly industrialized as England or Germany. A survey ordered by the government showed, however, that there were throughout the country hundreds, even thousands, of small firms capable of producing war material or contributing to its production. These were pressed into service to supplement the efforts of government factories and great firms such as Schneider. Not all could start work immediately, because an unselective call-up had deprived many of their manpower and weeks or months elapsed before essential workers could be brought back from the trenches. The fact remains that France survived the perils of the next few years by mobilizing virtually the whole of her industrial resources, including those of the light engineering industry and much of the textile industry, in support of her war-effort. Her military leaders were, however, so slow to abandon their faith in frontal assaults that much of the ammunition produced was squandered in useless offensives whose cost in killed and wounded vastly aggravated the nation's manpower problems.

So far as her army and her arms factories were concerned, Germany began the war rather better equipped than France for the kind of war she had to fight. Her strategists did not foresee that Moltke's attempt to defeat the French in six weeks would be followed by a long period of siege warfare, with both sides manning continuous lines of trenches. But they did foresee that heavy guns might be needed to deal with the French frontier fortresses and with defended places such as Lille, and in consequence their troops were fairly well provided with such weapons. Because the Germans had made a close study of the South African, Russo-Japanese and Balkan wars and had drawn conclusions from which the French recoiled, they were also well provided with machine-guns, trench mortars, barbed wire and entrenching tools. And the German Army, like the British, had adopted well before the outbreak of war the sensible habit of dressing its combatant troops in inconspicuous uniforms.

Even so, Germany did not reckon before 1915 on a long war. This was well shown by her failure to safeguard her external sources of supply. At the cost of making an enemy of Britain, she provided herself before the outbreak of war with a powerful High Seas Fleet. Yet she made no serious attempt to use her naval strength to protect her trade routes or her communications with possessions and interests in Africa, the Far East and the Pacific. Early in the war, all her colonies and concessions overseas either passed into the hands of her enemies or were cut off by Allied command of the sea. Throughout the remainder of the war she depended for raw materials on her own resources, on imports from countries accessible by land or by short sea routes beyond the reach of Allied warships, on occasional cargoes which slipped through the Allied blockade in merchant vessels or were carried by submarine.

A notorious example of a cargo which slipped through the Allied blockade was a consignment of nickel carried from New Caledonia to Norway early in the war in a Norwegian merchant vessel. After the ship had sailed, the French colonial authorities in New Caledonia warned Paris that they had reason to believe that the nickel was intended for ultimate delivery to Friedrich Krupp AG and had been bought and paid for by that company. The ship was intercepted by a French cruiser and forced to put into Brest. But the government, unwilling to forfeit the goodwill of Norway, decided that she should be allowed to proceed with her cargo to a Norwegian port. The nickel was duly delivered to the Ruhr about three weeks after her departure from Brest.[3]

Possibly the British might not have been so lenient had the decision rested with them alone. Their policy until the United States entered the war in 1917 was to impose the most rigorous blockade they could without incurring the active hostility of the Americans and to gain, as far as was possible, a monopoly of overseas supplies of scarce or crucial raw materials by pre-emptive buying. Copper was one of the most important of these materials. The Germans succeeded in obtaining a certain amount of copper from overseas through neutral countries, but deliveries were few and far between. By the spring of 1915 the British government controlled 95 per cent of America's exportable surplus of copper and was thus in a position to insist that exports went only to customers not on its blacklist.

The British would not, of course, have been able to take such a tough line had they not been rich and powerful. Britain was the world's leading naval and mercantile power, British finance houses and British nationals held massive investments in North and South American commercial enterprises and public utility undertakings, and roughly three-quarters of Britain's overseas trade was with countries outside the British Empire. Even so, the government's decision to enter the war on the side of France and Russia confronted her statesmen and officials with formidable problems of procurement and supply. Austria, France, Germany, Italy and Russia, besides being naval powers, possessed huge standing armies backed by large numbers of reservists who had undergone long periods of compulsory military training and were still of military age. Britain had only a small professional army, supplemented by a Territorial Army of volunteers who had received only intermittent training and a small Special Reserve intended to provide drafts to replace casualties. The War Office was not required before the outbreak of war to prepare for active service more than a tiny Expeditionary Force of six divisions of all arms and one cavalry division. Britain's arms industry had been kept in being for the best part of the past forty years by orders from foreign governments and the Admiralty, eked out by very small orders for guns and shells for the army.

Furthermore, on the outbreak of hostilities Britain had no full-time Secretary of State for War. The last incumbent had resigned some months earlier in consequence of a dispute about the army's role in the event of civil war in Ireland. Since that time the Liberal Prime Minister, Herbert Henry Asquith, had looked after the War Office on a part-time basis. When war came, Asquith broke with tradition by conferring the post on a professional soldier, Field-Marshal Lord Kitchener.

Kitchener had spent much of his service career in Egypt or India. He had little knowledge or understanding of conditions at home, the duties of a Cabinet Minister in a government committed to the principle of collective responsibility, or the limitations of an arms industry geared to peacetime requirements. He predicted that because the French were notoriously inefficient, or alternatively because the Germans were a brave people who would not readily admit defeat, the war would not be over in a few months but would last three years or more. When an appeal for volunteers to keep the Expeditionary Force up to strength brought not the hundred thousand recruits expected but many times that number, he made it his business to raise an army very much larger than any the government had hitherto undertaken to put into the field. The problem of equipping scores of new divisions was thus superimposed on that of remedying the deficiencies of the existing Expeditionary Force.

Kitchener did his best to cope with this enormous task, but he was handicapped by his inexperience, distrust of civilians and innate conservatism. When he received insistent demands from the Expeditionary Force for more ammunition and especially for more shells filled with high explosive, he did not show these appeals to his ministerial colleagues but contented himself with putting pressure on subordinates. The Cabinet as a whole received so little information from Kitchener about the situation in France that on 20 April 1915 Asquith publicly denied that the Expeditionary Force was short of ammunition. Less than three weeks later, British troops launched an attack at Festubert after a bombardment which their lack of high explosive made so ineffective that in some places the enemy's wire was still intact when the infantry went forward.

Meanwhile the leading armament firms were deluged with orders out of all proportion to any received from the Ordnance Department in recent years. As soon as hostilities began the Master-General of the Ordnance and his staff, foreseeing that a handful of government factories would not be able to meet all the demands of even a short war, made haste to address themselves to the relatively few commercial firms with whom they had dealt in peacetime. They made little attempt to tap the resources of the engineering industry as a whole. Manufacturers not on their list who offered their services were treated almost as suppliants. Those who persisted in face of initial discouragement were shown, at government factories, the kind of work they would have to tackle if they did have the temerity to enter the arms industry. If their willingness to serve survived this experience, they were grudgingly invited to apply to the department's approved suppliers for sub-contracts. Since the approved suppliers were overwhelmed with work, did not know how long the war would last and were understandably reluctant to disclose their secrets to firms which might become competitors when peace returned, this procedure did not lead in the early stages of the war to any widespread adoption of the sub-contracting system.

Nor was it meant to do so. The Master-General of the Ordnance and his subordinates were openly distrustful of sub-contracting. Their distaste for it was exceeded only by their reluctance to deal with firms they did not know.

An important consequence of this state of affairs was that unrealistically large orders for shells were placed with armament firms uniquely qualified to make guns but little better equipped than many general engineering firms for the manufacture of shell-cases in vast quantities. Between the outbreak of war and the end of 1914 the Ordnance Department ordered about six million shells from the leading armament firms in the United Kingdom, fewer than a million from government factories, about three million from suppliers in Canada and the United States. If the order for six million shells had been spread over a much larger number of firms, doubtless some firms would have found the work beyond their scope but probably most of the shells would have been delivered on time. As things were, only about a third were forthcoming by the scheduled date. The armament firms were afterwards reproached by government spokesmen with promising more than they could perform, but little was said about the pressures to which they were subjected. The usual procedure in the early months of the war was that each firm was told to produce so many guns and so many shells of a particular type by a given date. Knowing they would be eyed askance if they raised objections, the firms seldom questioned these edicts until the government's demands became so manifestly unreasonable that they were forced to do so. In the meantime they assumed that somehow or other the additional labour needed to increase their output to the extent required by the authorities would be made available.

What in fact happened was that the government provided capital to enable the firms to make additions to their plant where this was needed, but did nothing in the early part of the war to ease their manpower problems. On the contrary, Kitchener's recruiting drive enticed many thousands of workers from the arms factories to the parade ground and ultimately to the trenches. Vickers, for example, lost a thousand young men from their Sheffield works alone.[4]

Some firms were reproached not only with promising more than they could perform but also with accepting orders from Allied governments which they must have known they would not be able to fulfil if they were to meet their own government's requirements. This complaint had some substance; it would have had more if the government had been able to give the industry a reliable estimate of its needs and a clear definition of the circumstances in which foreign orders could be accepted. As it was, the firms received little more than vague instructions to accept orders from abroad only if they could be met without prejudice to requirements which changed with bewildering rapidity.

Armstrongs, for example, were large contractors to the Admiralty, but for some years before the war had received only small orders from the Ordnance Department because 80 per cent of the guns and 77 per cent of the shells supplied to the army in peacetime came from government factories. On 18 August 1914, they were asked to provide 162,000 shells for 18-pounders, at the end of the month nearly twice as many. In the following October they received an order for 400,000 shells to be delivered on the same date as those ordered on 18 August. Immediately after the Battle of Mons the Ordnance Department ordered seventy-eight 18-pounder guns, but the figure was raised by stages during the next six weeks to a total of 460.

Other firms had similar experiences. Vickers had sold about a hundred machine-guns to the Ordnance Department between 1905 and the outbreak of war; on 11 August 1914, they received an order for 192. On 10 September they were asked for another hundred; on 19 September for ten times as many; a few days later for five hundred more. Like Armstrongs, they were told after the Battle of Mons to supply seventy-eight 18-pounders; by the end of the first week in October the number ordered from them rose to 360. In addition, they were called upon in the early months of the war to produce large numbers of shells, to build at first sixteen and later thirty-two 9·2-inch guns, and to design and manufacture an entirely new howitzer.

On 21 December 1914, the armament firms at last rebelled. At a meeting with representatives of the War Office on that day, they made it clear that they would accept no more orders necessitating additions to their plant unless more labour was forthcoming. Sir Trevor Dawson of Vickers asked that recruiting offices should be told to accept no more volunteers from the arms industry. He also suggested that workers should be drafted to the industry from firms not engaged in war production.

An attempt was made to give effect to the second of these proposals, but without success. General engineering firms, when asked to part with workers for the benefit of the armament firms, protested that their own factories were fully capable of producing munitions of war. The military authorities then belatedly agreed that firms not on the approved list should be allowed to apply directly for army contracts, with the proviso that their works should first be inspected and pronounced suitable. The Board of Trade thereupon proposed a comprehensive survey of general engineering firms, but the scheme broke down because the Engineering Employers' Federation was not satisfied with the information provided by the authorities about the nature of the contracts they were prepared to offer.

Shells were not the only bottleneck. Supplies of propellants and other explosives in Britain on the outbreak of war were far from satisfactory. Cordite, the standard British smokeless propellant, was an article of commerce used for sporting guns as well as weapons of war. Waltham Abbey, the only government factory that produced it, was incapable of meeting all the needs of the armed forces in time of war, but commercial manufacturers were more than willing to fill the gap as long as the necessary raw materials were forthcoming. The most important of these was acetone, commonly made from wood. A factory in the Forest of Dean established by the Department of Woods and Forests before the outbreak of war was intended to meet Waltham Abbey's requirements. There was also a small commercial factory in East Anglia which aimed at making acetone from potatoes. Most commercial users of acetone depended, however, largely on imports from the United States. Despite the establishment in the early months of the war of new acetone factories in Devonshire, Scotland and the New Forest, the country was threatened with a serious shortage when American suppliers were found to have sold more acetone for future delivery than they could produce. At the government's request Dr Chaim Weizmann of the University of Manchester—afterwards well known as a leader of Zionism—came to the rescue by devising a commercially viable method of making acetone from the starch content of maize and other cereals.

So far as explosives other than propellants were concerned, the crucial factor was the high explosive needed for the shells demanded by the Expeditionary Force to reduce the enemy's wired and entrenched positions, gun emplacements and machine-gun posts. The enormous difficulty experienced by the authorities in meeting these demands was due partly to the fact that, barely three months before the war began, they had decided to replace lyddite as their standard high explosive by trinitrotoluol, better known as TNT. By August 1914 production of lyddite had almost ceased, but production of TNT had not yet got into its stride. The sole source was a single factory whose output a month after the outbreak of war was less than twenty tons a week.

The situation was further complicated by the international character of the explosives industry. By far the largest producer of explosives in Britain was Nobel's Explosives Company, formed under another name in 1871 to exploit patents held by Alfred Nobel, the inventor of dynamite. This was not a wholly British concern. It was controlled by a holding company, the Nobel-Dynamite Trust Company, which also controlled companies established by Nobel in Germany and Austria, among other countries. Another holding company, the Société Centrale de Dynamite, controlled companies in France, Italy, Switzerland and Spain. Although Nobel's original patents had expired, between them the two holding companies controlled more than nine-tenths of the world's trade in explosives. Most of their revenue came from the sale of explosives used in mines and quarries. But they were also interested, through their subsidiary and associated companies, in the production and distribution of military explosives.

On the outbreak of war the directors of Nobel's Explosives Company and its German and Austrian counterparts saw that endless difficulties would arise unless the assets controlled by the Nobel-Dynamite Trust Company were divided along national lines. They were allowed by their respective governments to meet in neutral Holland to discuss the matter. The outcome was that Nobel's Explosives Company became an all-British company, British shareholders in the Nobel-Dynamite Trust Company were allotted shares in it, and the trust company was wound up. Similarly, German and Austrian shareholders in the trust company exchanged their holdings for shares in companies registered and domiciled in Germany or Austria.

This reform had, of course, no direct effect on production. To meet wartime needs, the British government was compelled to set up national explosives factories which drew on the knowledge and experience of commercial producers. At the same time the military authorities were persuaded, with some difficulty, to substitute for pure TNT, as a filling for shells, a mixture of TNT and ammonium nitrate to which technical objections were made but which had the advantage of giving a bigger bang for less money. The French were better placed than the British to concentrate after the outbreak of war on the production of shells filled with high explosive, but they, too, had great difficulty in matching Germany's output. As late as the summer of 1915, the Germans were turning out at least two shells for every one produced by the British and the French between them.

Ironically, the member of Asquith's government who did most to awaken his colleagues to the necessity of adopting realistic measures to speed the flow of munitions was not the warlord Kitchener but the anti-militarist, almost pacifist Chancellor of the Exchequer, David Lloyd George. Lloyd George proposed in the second month of the war that a Cabinet Committee should be appointed to look into the whole question of the supply of guns, shells and rifles. Kitchener resisted the proposal until October. A committee with Kitchener himself as Chairman was then appointed. The outcome of six meetings held before it was disbanded early in 1915 was that Armstrongs, Vickers, the Coventry Ordnance Works and William Beardmore and Company were urged to increase their output on the understanding that the government would meet the cost of additions to their plant, and that large orders for rifles and explosives were placed in the United States. But these decisions did not enable the armament firms to overcome their labour problems, and no large deliveries of rifles or explosives from the United States were expected in the immediate future.

In the following March the government took power under an amendment to the Defence of the Realm Act to assume control of any factory or workshop, requisition empty buildings to house war workers, and annul contracts that might prevent general engineering firms from switching to war production. Lloyd George pressed for the appointment of a Munitions of War Committee with executive authority, but the body set up in response to his demands was unable to extract from the military authorities any reliable estimate of future needs or persuade them to relinquish any of their power. The creation of the committee did, however, amount to an admission by the government that the procurement and supply of arms in a major war were tasks beyond the unaided capacity of the War Office and the Admiralty.

In May the Liberal government, already discredited by its manifest inability to furnish the Expeditionary Force with enough ammunition for a sustained offensive, was swept from office by the backwash of a dispute between the First Lord of the Admiralty and the First Sea Lord about the naval expedition to the Dardanelles. Lloyd George became Minister of Munitions in the coalition government Asquith was then obliged to form. He assembled a team of what he called 'captains of industry' to organize the production of weapons and munitions of war throughout the country.

However, captains of industry were not necessarily industrialists. Lloyd George's henchmen included civil servants as well as men whose knowledge of commerce and industry had been gained in callings as diverse as publishing, distilling, transportation, accountancy, arms manufacture and marine, civil, mechanical and general engineering. Many were lent to the government by their employers, who continued to pay their salaries for the duration of the war.

The basis of Lloyd George's system was decentralization. Outside London, the Ministry of Munitions established Area Offices at Newcastle, Manchester, Leeds, Birmingham, Cardiff, Bristol, Edinburgh, Glasgow, Dublin and Belfast. The Area Offices exercised powers of general supervision over nearly fifty Boards of Management, each of which was made responsible for furnishing a specified quota of war material. The Boards of Management looked both to commercial firms and their subcontractors and to government factories to fill their quotas. Altogether some seventy such factories—sometimes called national factories to distinguish them from the older-established government factories at Woolwich, Waltham Abbey, Enfield Lock and Farnborough—were set up by the end of 1915. They were the first instalment of well over two hundred established by the end of the war. Their products ranged from guns, shells and explosives to optical equipment, machine-tools, gauges and ammunition-boxes.[5]

As Minister of Munitions and later as Prime Minister Lloyd George did not aim, however, at driving private arms manufacturers out of business. The leading commercial firms remained the chief producers of most of the more important weapons and acted as sponsors or managers of many of the national factories. They also continued to turn out very large numbers of shells. Of the many millions of shell cases manufactured in the United Kingdom during the war, not far short of half came from firms working for the government on a contract basis, about a quarter from national factories supervised by such firms. The rest—about twenty-seven per cent—were produced by groups of national factories organized by Boards of Management on a co-operative basis.

The commercial firms also made important contributions to design and development. The government agency responsible on the outbreak of war for sponsoring new or improved weapons for both the army and the navy was the research department of the Royal Arsenal at Woolwich. In the summer of 1915, however, the Admiralty set up a research department of its own. The Ministry of Munitions established a similar department soon after its creation, but responsibility for designs, patterns, specifications and testing remained, so far as new or improved weapons for the army were concerned, for many months a bone of contention between the Ministry and the Ordnance Department. Eventually the question was settled in favour of the Ministry. The Ministry then assumed control of the research department at Woolwich and other research establishments hitherto administered by the War Office. The practical effect of these arrangements was that the Ministry of Munitions and the Admiralty provided encouragement, technical advice and sometimes financial support, but that most of the new or improved weapons adopted during the war originated in the drawing offices of commercial firms or government factories. Two outstandingly successful weapons, the Stokes mortar and the tank, were exceptions.

The Stokes mortar was invented by Wilfred Stokes, a member of an Ipswich family firm well known for its agricultural machinery and kindred products. The military authorities were lukewarm about it until its success was proved, but Lloyd George provided financial support, in the shape of an order for a thousand mortars and a hundred thousand rounds of ammunition, by drawing on a fund placed at his disposal by an Indian potentate.

The tank, or 'land ship', was born of the enthusiasm of a small band of soldiers, sailors and marines. One of them, Colonel Ernest Swinton, suggested to a director of Armstrongs in 1915 that the firm should develop the weapon as a private venture, but nothing came of the suggestion, which was made more or less on the spur of the moment and was not followed by any formal proposals. In the outcome the tank was developed by the Admiralty, but orders for production models were placed by the Ministry of Munitions. The leading armament firms did not contribute to the designs of the hulls or engines of the early tanks, but Armstrongs designed the first successful guns and gun-mountings for them, and Vickers were suppliers of materials and armour.

The fact remains that in Britain, as elsewhere, private arms manufacturers were able to maintain their identity only by becoming partners of the government in a joint enterprise. The firms were not nationalized in the sense that shareholders were required to surrender their holdings to the state. Government spokesmen expressly disclaimed any intention on the part of the authorities of telling them how they should manage their affairs. Nevertheless it was, for all practical purposes, the government that decided how their assets should be employed. If the government wanted shells but had no immediate use for armour, then Armstrongs and Vickers made shells at the cost of allowing their expensive plant for the production of armour to remain idle.

The government was also in a position to exert pressure on the arms trade by controlling prices. Apart from special powers conferred by wartime legislation, the authorities had always possessed the right to requisition goods for the armed forces. In principle, these had to be paid for at rates determined by the 'fair market value'. But this term became meaningless when the Admiralty and the Ministry of Munitions were the sole buyers. The only sound method of deciding how much should be paid for any article in such conditions was to determine the cost of production and add a reasonable margin to cover overheads and profit.

This was, of course, the method normally used by commercial firms to calculate the lowest price at which they could afford to sell a given product. In the early stages of the war, the government's suppliers of shell charged prices arrived at in this way, each firm using its own methods of cost accounting and making its own assessment of the proportion of its overheads an order could reasonably be made to bear. One consequence was that prices differed a good deal from firm to firm. But that—as the Admiralty recognized and the Ministry of Munitions rather grudgingly admitted—did not necessarily mean that the prices charged by some firms were excessive. Costs of production, and therefore prices, were bound to differ where the exigencies of war compelled firms with various backgrounds to undertake work which some were better qualified to do than others.

Even so, a difference of nearly forty per cent between the prices charged by two contractors for the same article was bound to excite remark. Not surprisingly, one of the first steps taken by the Ministry of Munitions after its creation in the summer of 1915 was to investigate costs of production in the light of estimates by its engineering experts and inspection of contractors' books. In the following December the Ministry gave five of its principal suppliers notice of the termination of their current shell contracts and announced that prices were to be reconsidered. The firms were then presented with a list of prices described by the Ministry as based on ascertained costs. Some hard bargaining followed before agreement was reached on a new scale of charges which were not in all cases lower, and in one or two instances were actually higher, than those hitherto in force.

Finally, it was the government that decided how much of the money paid to them the firms should be allowed to keep. Their profits were restricted both by legislation which applied only to profits derived from the manufacture of armaments, and by a tax levied on all manufacturers whose annual profits during the latter part of the war exceeded those for the fiscal year 1914–15 by more than a hundred pounds.

The circumstances in which wartime governments came to accept the principle that special restrictions should be imposed on profits derived from the manufacture of armaments arose from attempts made by the Liberal government in the late winter and early spring of 1915 to put an end to strikes and restrictive practices in the arms industry without resorting to compulsion. When the amendment to the Defence of the Realm Act adopted in March was first considered, clauses which would have made strikes illegal were rejected in favour of less drastic measures. Representatives of the trade unions were then invited to confer with the Chancellor of the Exchequer and the President of the Board of Trade in a room at the Treasury whose most prominent feature was a gilt throne last occupied by George I. They agreed that, as long as the war lasted, unskilled or semi-skilled workers in arms factories should be allowed to do jobs normally reserved for skilled men, but only on condition that they were paid the usual rate for the job and that steps were taken to restrict profits. Representatives of both sides then signed a document embodying these terms.

This exchange of pledges—known thereafter as 'the Treasury Agreement'—seems to have been regarded by everyone who attended the meeting as fair, just and timely. But trade unionists did not surrender their belief that the dilution of labour, even though it might have to be accepted in time of war, was wrong in principle.

However, whether right or wrong, clearly the dilution of labour was highly necessary. Between the outbreak of war and the autumn of 1915 about a quarter of a million men left heavy industry and its ancillary trades for the armed forces. Only about five thousand were brought back in accordance with a scheme which permitted the release of skilled men urgently needed by firms engaged in work of national importance.[6]

Meanwhile strikes were frequent. Some were clear infractions of the Treasury Agreement, but were defended by representatives of organized labour on the ground that the government had not yet redeemed its promise to restrict profits.

The reason was not that the government had taken no action in the matter. Armstrongs and Vickers were invited in March 1915 to submit proposals to the Board of Trade for the limitation of dividends during the rest of the war and for a short period after it. On these the government intended to base a scheme which would be applied to the whole of the arms industry. However, in the course of the ensuing discussion it was suggested that divisible profits rather than dividends should be restricted. A mutually acceptable method of assessing and restricting divisible profits had then to be arrived at by negotiation. The result was that it was not until June, when the Liberal government had fallen, that the coalition government then in office was ready to assume statutory powers to limit the wartime profits of the arms industry. These powers were conferred by the Munitions of War Act, which also empowered the government to prohibit strikes, lock-outs and restrictive practices, settle industrial disputes by compulsory arbitration, and enrol volunteers willing to undertake war work wherever their services might be needed. The Finance (No 2) Act of 1915 completed the machinery of limitation by imposing a special tax on all wartime profits which exceeded taxable profits for the last fiscal year by the stipulated margin.

In the outcome, the wartime profits of the leading British armament firms were not only substantially lower than those for which their representatives stood out during the negotiations in

1915, but also slightly lower than those first proposed by the Board of Trade. In the last full year of peace Vickers had made a profit, after taxation, of £911,996 on an issued capital of just under six million pounds. Armstrongs had made slightly less. Walter Runciman, President of the Board of Trade, calculated that the two firms ought to be allowed to make in each of the war years up to £1,150,000 and £960,000 respectively, but afterwards conceded maxima which exceeded these limits by some five per cent. In practice, even the lower figures were not quite attained. During the five years from 1914 to 1918 Vickers made an average annual profit, after taxation, of £1,120,488 on an issued capital which rose by the end of 1918 to close on nine million pounds. Armstrongs' wartime profits bore a very similar relationship to Runciman's figure.

Thus there is no ground, so far as the leading British firms were concerned, for the popular belief that the huge demand for guns and shells created by four years of trench warfare enabled arms manufacturers to wax fat on swollen profits. Ordinary shareholders in Vickers received throughout the war the same 12½ per cent dividend as was paid in 1913. Higher dividends had been paid on many occasions between 1882 and 1907, when the country was at peace.

Nor were British arms manufacturers able to pile up huge reserves. Shortages of staff arising from the demands of war caused the accounts department of Vickers to fall so far behind with its work that divisible profits for the period from 1916 to 1919 could not be finally assessed until after the peace treaties were signed. They were then lumped together at £6,612,439 for the four years. Dividends, including an ordinary dividend of 11¼ per cent for 1919, absorbed £5,488,885, leaving £500,000 to be added to the company's general reserve and £623,554 to be carried forward. These were substantial sums, but they were not unduly large for a company which equipped itself for the switch from war to peace by undertaking in 1919 a capital expenditure of nearly thirteen million pounds, financed by new issues of ordinary and preference shares.

British armament firms did, however, have one substantial advantage over their counterparts in Continental Europe. In all the countries which found themselves committed to war in 1914, arms manufacturers received large government orders which went at least some way to compensate those accustomed to sell their products abroad, for the loss of foreign markets. In all of them, divisible profits were restricted to some extent by government action of one kind or another. But only in Britain did wartime governments make serious, sustained, and on the whole successful efforts to maintain the real value of the national currency. The result was that, while most European firms were haunted almost from the outset by the fear that inflation might reduce the purchasing power of their liquid assets almost to vanishing point by the time the war was over, the British never doubted after the first few weeks of hostilities that their country's economic and financial structure could withstand the stresses of even a long war. Vickers continued to invest in government securities even when they were borrowing large sums from the government and from their bankers to finance work in progress and stockpiling. Air raids on London caused some absenteeism and loss of production at Woolwich Arsenal. U-boat attacks on merchant shipping brought the British Isles almost within sight of starvation. Despite the requisitioning of dollar investments held by British subjects domiciled in the United Kingdom, Britain's reserves of gold and dollars came so close to exhaustion that large loans had to be raised in the United States to finance purchases of raw materials, foodstuffs and manufactured goods. Through all these trials, a serious decline in the value of the pound sterling in the international money market remained unthinkable.

No such confidence in the staying power of national currencies was felt in Paris, Rome, Berlin or Vienna. The French, with their most productive ironfields and some of their richest provinces in German hands, had to import a great deal that was normally produced at home and became dependent on the British to finance their American purchases. Italy's decision to throw in her lot with the Entente Powers destroyed her trade with Central Europe and forced her to rely largely on her allies for the means of keeping her armies in the field. Germany, dragging a reluctant Austria in her wake, adopted such desperate expedients to sustain her war effort that her statesmen were unable to contemplate any end to the war but outright victory, for fear that any other outcome might prove financially, economically and morally disastrous. Theobald von Bethmann-Hollweg, German Chancellor until pressure from the army led to his dismissal, admitted that Germany's invasion of Belgium was a crime, her requisitioning of Belgium's industrial resources 'an act of piracy'. The sinking of British, Allied and neutral merchant vessels by U-boats whose commanders made no provision for the rescue of passengers and crews was a flagrant violation of international law. Not only Germany's rulers but also her leading industrialists incurred lasting odium by decreeing or acquiescing in the employment of Belgian conscripts in German factories. Not surprisingly, the Germans did not feel until their armies were on the verge of defeat that they could afford to make peace except on their own terms.

Even so, events falsified the prediction that economic factors would make it impossible for Europe's leading industrial powers to sustain the burden of a long war. France and Britain, recovering from a shaky start, succeeded with the help of American productive capacity and financial support in providing themselves by 1918 with such vast reserves of guns, shells, tanks and aircraft that, after the failure of the German offensives in the spring and early summer, they faced with confidence the prospect of another year of war. Italy, surviving a tremendous setback at Caporetto, managed with British and French help to keep her armed forces in being as long as the war lasted. The Central Powers were not prevented by the Allied blockade from maintaining an output of weapons and ammunition which, at any rate until the war was three parts over, more than matched that of their opponents.

Russia was in a very different position. In the course of peacetime conversations with the British and the French, Russian officers spoke of mobilizing an army of six and a half million men. But their hopes of equipping the five million reservists they counted on adding to their standing army were founded on the assumption that an arms industry revitalized with British and French help would be in full production by the time war came, or that if it were not they would be able to buy everything they needed from Vickers and Schneider. As things turned out, the events of 1914 found Russia unprepared and Britain and France with virtually no exportable surpluses of arms and ammunition. On the outbreak of war the Russian Army had only meagre reserves of shells for its artillery, was desperately short of rifles and did not possess adequate supplies of even such humble necessities of war as cartridges, boots, warm clothing, wire-cutters and entrenching tools. Six months later, Russia's entire output of rifles was only about 40,000 a month. In some sectors, fewer than a third of the recruits who joined combatant units could take their places in the firing line as soon as they arrived; the rest had to wait until they could be issued with rifles recovered from the dead or wounded.

General Vladimir Sukhomlinov, the Russian Minister of War, attributed this lamentable state of affairs to Russia's habit of relying in past years on foreign suppliers and the consequent discouragement of producers at home. His critics pointed out that Sukhomlinov himself was largely responsible for his country's failure to organize arms production on an adequate scale, was notoriously indolent and corrupt and was strongly suspected of sympathizing with the Central Powers. (He was afterwards found guilty of 'abuse of power' but acquitted of treason.) In any case, Russia differed from France and Britain in not possessing a well-developed general engineering industry which could be readily mobilized in support of her war effort. Furthermore, the facts of geography were against her. From the start of the war the seaward approaches to her Baltic ports were dominated by German warships. Before long she found herself in a position which gave her no choice but to deprive herself of access to the Mediterranean by declaring war on Turkey. Hitherto the world's largest exporter of wheat, she sacrificed by this gesture about nine-tenths of her foreign trade. A consequent rise in the price of wheat increased the profits of North American farmers and speculators but was highly detrimental to the British, who relied on imports for four-fifths of their wheat and wheat-products, including nearly all the hard-grain wheat from which the flour used in commercial bakeries was milled. One of the arguments adduced in favour of the Allied expedition to Gallipoli in 1915 was that the capture of Constantinople, by reopening the Straits to grain-ships outward bound from Odessa, would bring down the price of bread.

An even more serious disadvantage of the closing of the Straits was that Russia was left with Archangel, Vladivostok and Murmansk as her sole ports of entry for arms, ammunition and raw materials shipped from abroad. The first was icebound throughout the long northern winter, the second was thousands of miles from European Russia, the third was separated from the chief centres of production and distribution by vast tracts of lake and forest traversed only by a railway still incomplete when hostilities began. In 1916 and 1917 huge dumps of military stores awaiting distribution accumulated at all three places. These attracted scores of pillagers and provided a ready source of supply for saboteurs and insurgents in search of weapons, cartridges and explosives.

Lloyd George suggested after the war that the British and the French, instead of wasting enormous numbers of shells on futile offensives at Loos and in Champagne, would have been well advised to concentrate in 1915 on saving the Russians from being driven back in Poland. This argument assumes that the effort which produced the shells wasted at Loos and in Champagne could equally well have produced the shells, rifles and cartridges needed by the Russians. Doubtless British and French arms manufacturers could, if they had been told to do so, have completed by the late summer a considerable output of the appropriate weapons and ammunition. Whether this material could have been shipped to Russia and distributed to combatant units in time to stem the German advance in Poland is another matter. The poor carrying capacity of the sparse Russian rail network ranked with the all-pervading inefficiency and venality of Russian administrators among the chief causes of a state of affairs which was soon to make the fall of the Tsarist régime inevitable.

SEVEN

American Intervention and the Armistice

On the outbreak of the First World War the United States possessed a substantial navy but only a very small professional army. This was intended as a focus of strategic doctrine and a nucleus round which a citizen army would assemble in time of need. American defence thinking rested on two assumptions. The first was that, in the event of war, the navy and small local garrisons would suffice to protect the Continental United States and its overseas possessions until a citizen army was ready to take the field. The second was that a nation of pioneers accustomed to the use of arms would need comparatively little military training.

Neither of these assumptions was sound. By 1914 the United States had long ceased to be exclusively, or even mainly, a nation of pioneers. Inhabitants of Texas and Wyoming might be well versed in the handling of firearms, but there was no reason to suppose that citizens of New York or Milwaukee would need less military training than French or German conscripts. And it was unrealistic to assume that the navy, supplemented by local garrisons, could defend the whole of the eastern seaboard and also provide adequate safeguards for the West Coast, the Hawaiian Islands and the Philippines. The opening of the Panama Canal in 1914 went some way towards reconciling the navy's commitments with the axiom that the fleet ought never to be divided. But a threat to American interests in the South-West Pacific would still be hard to counter if it coincided with trouble nearer home.

Since the North American sub-continent was not in any immediate danger, this did not matter very much. What did matter was that far more was needed to equip a citizen army for eventual participation in the kind of war that erupted in 1914 than a modest array of field artillery and a stock of rifles, carbines and pistols manufactured at leisure and stored in armouries until units were ready to take them up. Armies had come to depend to an extent scarcely foreseeable when the Spanish War was fought on elaborate communications networks, on vast accumulations of wheeled transport, above all on ample fire-power. An American expeditionary force capable of making an effective contribution to the war in Europe would need large numbers of modern field, medium and heavy guns, still larger numbers of mortars and machine-guns. Such weapons could not be produced in useful quantities in a few weeks, or even a few months. Some of them could not be produced at all until designs were approved, prototypes tested, factories built or enlarged and provided with tools and plant. The military authorities could not hope to have them ready for use in an emergency unless crucial decisions were made long before the emergency arose.

The United States was a highly industrialized nation. It was also a peace-loving democracy in which the army was kept firmly in its place. Companies and corporations whose arms-producing potential were commensurate with the army's needs were not lacking, but their relations with the military authorities were very different from those subsisting between arms manufacturers and ordnance departments in European countries. They received no contributions from the Department of the Army to the cost of maintaining research and development departments, little or no indirect support in the shape of reliable indications of the nature of the weapons it might be worth their while to develop as private ventures in the hope that, if they were successful, the army would adopt them. An American inventor of a weapon with a purely military application could not expect to receive a penny from his own government until he perfected his design and the prototype was tested and approved. He might then be awarded a contract if he could find a manufacturer able and willing to undertake quantity production on acceptable terms. Not surprisingly, such pioneers of the automatic weapon as Hiram Maxim, John Browning, Isaac Lewis and Benjamin Hotchkiss made arrangements for their inventions to be manufactured and marketed by foreign firms. New field, medium and heavy guns were still less likely to be developed for the United States Army by private venturers.

The army did, of course, possess facilities of its own for research and development, but at best these could not do much more than enable experts to keep track of current trends. There is a big difference between knowledge of the kind of equipment that may be useful in a future war and the ability to produce it in large quantities at short notice. Executive decisions and productive capacity are needed to translate theory into practice.

The disadvantages of reliance on a very small professional army and a citizen army not permanently embodied and not backed by a productive capacity commensurate with needs which may arise with disconcerting suddenness were apparent to the few people who bothered their heads about such things long before the outbreak of the First World War. A number of attempts were made between the end of the Civil War and 1914 to persuade the federal authorities to put the national defences on a sounder footing. All were defeated by public indifference and presidential reluctance to contravene the sacred principle that standing armies were dangerous to liberty. A growing hostility to Big Business in the shape of trusts and combines was an additional factor.

Soon after the outbreak of the war in Europe, advocates of military preparedness returned to the charge by suggesting to President Woodrow Wilson that an attitude of strict neutrality towards the conflict between the Entente and the Central Powers need not preclude the United States from equipping herself to meet the situation that would arise if she were dragged into it against her will. They proposed that an early start should be made with the training of reservists and that the government should not wait until American involvement was imminent before making provisional arrangements for production of the artillery and other heavy equipment of an expeditionary force.

The President admitted in conversation with his confidential adviser, the Texan Colonel Edward Mandell House, that there was something to be said for these proposals. Nevertheless he refused to sanction measures which might seem to imply that the United States was actively preparing to join the belligerents. Early in the war he said privately that his personal sympathies were entirely with Germany's opponents. 'No one,' he declared, 'knowing the situation and Germany as I do, could be other than heart and soul with the Allies.'[1] Yet his advice to the great mass of his fellow-citizens was that they should try to be neutral in thought, word and deed. In his annual message to Congress on 8 December 1914, he described the struggle in Europe as 'a war with which we have nothing to do, whose causes cannot touch us.' He was confident that the country would always be ready to defend itself to the utmost, but how that was to be done he did not say. He did, however, make it clear that he did not intend to 'turn America into a military camp' and thought it neither right nor necessary that the United States should be able to put an army into the field at short notice.

The effects of the President's attitude on American industry were predictable. Arms manufacturers, it was clear, could not expect to do much business with the Department of the Army while the country was at peace. On the other hand, there was nothing to prevent them and other manufacturers from selling their products to overseas buyers able to pay for them and to take delivery. Industrial firms, some of them with little or no experience of the arms trade, accepted large orders from foreign governments for a wide range of military equipment. Gold flowed across the Atlantic to pay for American raw materials, foodstuffs and manufactured goods. The British government requisitioned foreign investments and turned them into dollars. Large loans, secured on British and French credit, were raised on the New York money market. No longer a debtor nation, largely dependent on British capital to finance her westward expansion and the provision of public utilities for her cities, the United States became the proud but slightly uneasy possessor of a huge share of the world's gold reserves. Meanwhile American manufacturers equipped themselves on a massive scale for the production of weapons and ammunition of European calibre and pattern.

If Woodrow Wilson really thought of the war as one with which the United States had nothing to do, he was wildly mistaken. Apart from the financial aspect, American interests were profoundly affected by Japanese participation, the Allied naval blockade and Germany's response to it. The seizure by Japan of Germany's colonial possessions in the Pacific north of the equator put Japanese forces across the routes between the Philippines and the Continental United States. Japanese demands on the Chinese Republic, although soon modified or withdrawn in deference to world opinion, suggested that Japan meant to flout American wishes by swallowing China as soon as she felt strong enough to do so. The Allied blockade did not prevent American exporters from selling their goods on favourable terms, but the use made by the Allies of their command of the sea to ensure that only customers of whom they approved were served caused a great deal of resentment and led to many complaints about the detention of mail and cargoes. American lives were lost in ships sunk by German submarines. A diplomatic indiscretion revealed a plot by German and Austrian emissaries in the United States to hamper the Allied war effort by sabotaging American factories. The Germans, not content with defying international law on the high seas and in Belgium, proposed an alliance with Mexico on the understanding that the Mexicans would receive financial support and be helped to recover Arizona, Texas and New Mexico in the event of war between the United States and the Central Powers. These and other side-effects of a war in which the United States played for two and a half years the role of a more or less passive onlooker made a great deal of work for the State Department and cost the President many anxious moments.

In 1916 the President sanctioned an ambitious programme of naval expansion. This was intended to give pause to the Japanese. But completion of the programme would put the United States in a strong position to challenge British naval supremacy unless Britain, too, made large additions to her battlefleet. Colonel House warned his chief that, in risking an arms race with the strongest naval power, he might be making the same mistake as Germany had made before the war. Wilson brushed the warning aside. He was no longer heart and soul with the Allies. After two years of blockade, he had come to regard as hypocritical Britain's claim to have entered the war because Belgium had been invaded. 'Let us build a navy bigger than hers,' he said, 'and do what we please.'[2]

The President's assessment of Britain's motives was not altogether sound. British statesmen may not have had quite as high a regard for the rights of small nations as they professed to have; but they did have a high regard for treaties. Treaties were the cement that held together a great part of the British Empire. Britain, when she went to war in 1914, followed her traditional policy of not allowing a first-class power to gain control of the Low Countries. But Asquith and his colleagues could scarcely have hoped to carry the country with them if the Foreign Secretary, Sir Edward Grey, had failed to convince the House of Commons that Germany's demands on Belgium were a gross violation of the treaty by which the leading European powers agreed to respect her neutrality.

However, the President did not build a bigger navy than Britain's. No ships of the 1916 programme had been completed when, in the spring of 1917, the German government's refusal to call off unrestricted U-boat warfare gave him no choice but to ask Congress to recognize that a state of war existed between the United States and Germany. The United States then joined Britain and France not as an ally, but as an 'associated power'.

The manifest reluctance, and even anguish, with which Wilson committed the country to war have been generally attributed to the dismay with which he viewed the collapse of his policy of staying out of the fighting at almost any cost. It is true that, having clung tenaciously to that policy for the best part of three years, he had good reason to feel upset when circumstances forced him to appear in the unfamiliar guise of warlord. But he had other than merely personal grounds for despondency. The truth was that, notwithstanding his brave words about spending the whole force of the nation to achieve victory, the country was in no position to make an independent contribution to the defeat of the Kaiser's armies. The United States Navy was ready to take a hand in the Allied blockade. The United States Army was emphatically not ready to play even a modest role in the great battles impending on the Western Front. It was small and widely dispersed. It was not organized in divisions. There were no divisional or corps staffs.[3] No expeditionary force had been formed for service in Europe, and very little had been done to provide the equipment an expeditionary force would need. The German Army had gone to war in 1914 with 12,500 machine-guns. Nearly three years later the United States Army had 440. Furthermore, the nation did not possess the means of carrying an expeditionary force of more than a few divisions to the theatre of war, or of supplying it when it got there.

When the President made his announcement to Congress on 2 April 1917, Admiral William S. Sims of the United States Navy was already on his way to Europe to tell the Allies what the navy could do. General John J. Pershing, Commander-in-Chief designate of a still hypothetical United States Expeditionary Force, followed with a small staff in May. After discussing some of his problems with British and French officers, he announced his intention of assembling a first wave of a million men in France by the early summer of the following year. Later the Americans spoke of sending twelve large divisions by June or July, and at one time they even contemplated sending twenty-four. Pershing hoped and expected to command a self-sufficient, all-American force, with its own sector of operations carved out of the Allied front, its own transport, supporting arms and services, its own lines of communication linking its rear and forward bases in Europe with its logistical base in the United States.[4]

However, the United States Expeditionary Force was not destined to be self-sufficient, and its build-up was at first extremely slow. Only 61,531 American troops reached France in the first six months of war, only about 300,000 in the first twelve months. (The British, starting like the Americans with a small professional army, had sent half a million men to the front within six months of declaring war, not far short of a million within a year.) To save shipping and ease problems of production and supply, Pershing was forced to rely on the French for nearly all his field guns, on the British for most of his heavier pieces. As for aircraft, the United States Army had only a very small air arm when, in the summer of 1917, responsibility for military aviation was taken out of the hands of the Signal Corps and an ambitious programme of expansion was authorized. Some thousands of British-designed De Havilland DH.4 light bombers, to be built under licence, were ordered from American aviation firms, and a consortium of motor-car manufacturers undertook to design and produce a suitable engine. At the end of the following February, however, there was still not a single aeroplane of American manufacture in Europe. Even when American-built bombers did arrive, Pershing still depended on British and French factories for his scouts and fighters.

Deliveries of small arms were more satisfactory. The standard American infantry weapon when the United States entered the war was the Springfield M.1903. This was an improved version of a ·30-inch rifle based on Mauser patents for which the government paid two hundred thousand dollars when the weapon was tested and approved in 1905. The Springfield M.1903 remained in service until 1929, but was supplemented in 1917 and 1918 by a ·30-inch version of the ·303-inch Mauser-type rifle which American small arms manufacturers had been making for the

British since 1914. More than two million of these M.1917 rifles were manufactured in some eighteen months by the Eddystone Rifle Plant of Pennsylvania, the Remington Arms and Union Metallic Cartridge Company of Ilion, N.Y., and the Winchester Repeating Arms Company of New Haven, Connecticut.

The most recalcitrant problems of all were those connected with the shipping of the expeditionary force and its supplies across the Atlantic. The carrying-capacity of the American merchant fleet in 1914 was only about an eighth of that of the British merchant navy. The situation in 1917 was rather better, but still far from satisfactory. The government requisitioned ships under construction for the British and suspended the 1916 capital ship programme so that shipyards could concentrate on the building of cargo vessels, destroyers and submarines. Even so, a calculation made some six months after the United States entered the war showed that the tonnage allotted by the Shipping Board to the military authorities would not suffice to maintain an expeditionary force of more than six to eight divisions, at the most.

In November 1917, representatives of the President, the armed forces, the Treasury, the War Trade Board, the Shipping Board, the War Industries Board and the United States Food Controller attended a conference in London with British ministers and officials. Somewhat hopefully, the Americans said that they intended to raise the strength of their forces in France to twenty-four divisions by the following midsummer, and could find half the shipping needed if the British would provide the other half. Lloyd George shook them by pointing out that three-fifths of the British merchant navy was already fully committed to war service for Britain and her allies.[5] As the result of a 60 per cent decline in French food production, the disruption of Italy's economy and the closing of Russian granaries to the Western Allies, British ships had to be used not only to feed and supply the United Kingdom and Britain's armed forces in foreign theatres, but also to carry food and raw materials to France and Italy.

The Americans then reverted to a target figure of twelve divisions by midsummer. At a conference in Paris at the end of November and the beginning of December, they proposed that these divisions should cross the Atlantic at the rate of two divisions a month and that half of them should travel in American and half in British ships.

The British accepted these proposals, but foresaw a major attempt by the Germans to force a decision on the Western Front before the Americans could reach it in strength. The arrival of two incompletely trained American divisions a month, they thought, would not do much to save the situation if the Allies had to withstand an all-out offensive in the spring. They offered, therefore, to carry not merely their share of the twelve divisions but an additional 150 infantry battalions if the Americans would allow them to be brigaded with Allied troops until American divisions were ready to take them up. To find the necessary shipping they would have to allow their stocks of food and raw materials in the United Kingdom to fall to dangerously low levels, but they were willing to put up with that in order to get American troops into action as fast as possible. They pointed out that the shipping needed for three whole divisions would suffice for the infantry of more than twelve. Moreover, battalions serving with experienced British or French troops could be made battle-worthy in less than half the time needed to train raw divisions.

The military authorities in Washington were inclined to accept this offer. Pershing was not. He did not want his battalions to be widely dispersed to suit the convenience of Allied commanders, or used to plug gaps in the Allied line. He wanted them to take their places in American brigades and divisions serving under American officers in an American sector. He foresaw that if the British, after carrying their share of the stipulated twelve divisions, used any further shipping they could spare to carry only infantrymen and machine-gunners, the formation of the rest of his divisions might be postponed for an indefinite period. As for the President, he was not prepared to say that the offer should be declined, but thought Pershing should be allowed to make such arrangements with the British as might seem best to him.

The outcome of a triangular debate between Pershing, the Allies and Washington was a compromise. Pershing's troops would be shipped to Europe by divisions. The infantry and ancillary troops would then train with Allied troops; commanders and staff officers would be attached to Allied formations to gain experience; gunners and airmen would learn to handle their equipment under their own officers, supplemented where technical guidance was needed by British or French instructors.

However, that did not end the matter. When the Germans

opened their spring offensive on 21 March 1918, Pershing had only one fully-trained division in the field. In addition, two divisions were going through the last stages of their training in quiet sectors of the French front, and one division was in reserve. The British addressed an urgent appeal to President Wilson, asking that all four divisions should assume an active role and that 120,000 infantrymen and machine-gunners should be sent from the United States to Europe in each of the next three months. Wilson agreed, after consulting the Department of the Army, to send 120,000 infantrymen and machine-gunners a month for three or even four months, but stipulated that Pershing should be free to allot them at his discretion to British, French or American divisions.

This concession brought an indignant protest from Pershing. Again the outcome was a compromise, and again the British went a long way to meet his wishes. In May and June, priority was to go to the infantry and machine-gunners of six divisions. Any further shipping that might be available would be used first to carry the remaining troops of those divisions. Only when these had been shipped would the infantry and machine-gunners of other divisions be embarked.

So much for theory. In practice, about three times as many American troops crossed the Atlantic between April and July as had made the voyage in the previous twelve months. Another six hundred thousand shipped after the end of July gave Pershing a total of 1,868,000 troops on his ration strength at the beginning of November. About 51 per cent of these troops were carried in British or British-controlled vessels, about 46 per cent in American, between two and three per cent in others.[6]

This total included a fairly high proportion of non-combatants and non-divisional details. Even so, if depot and replacement divisions are included, Pershing had by the end of the war the equivalent of forty-one exceptionally large divisions in major formations formed or forming. Two American divisions, each equivalent in rifle strength to two French divisions, joined the French on 18 July 1918 in opening the crucial attack which drove the Germans from their salient near Château-Thierry. More than twenty such divisions took part in the final Allied offensives between September and November.

Some months before the first of these attacks was launched, the

Allied and Associated Powers had agreed to put Foch in supreme command of their land forces on the Western Front. The use of American divisions in sectors held mainly by French or British troops was not hampered, therefore, by considerations which might otherwise have made Pershing reluctant to see his forces employed in mixed formations. Formations used by Foch towards the end of the war included an American corps of American, British and French divisions supported by American, British and French air forces.

Not surprisingly, one of the lessons drawn by American strategists from experience in the First World War was the importance, where troops of more than one nation fought together, of subordinating 'national jealousies and suspicions and susceptibilities of national temperament' to the purpose in view by the adoption of a system of unified control, and if necessary by the appointment of a supreme commander served by an integrated staff.* This lesson made so deep and lasting an impression that one of the first things the Americans did when they found themselves at war in 1941 was to ask that the British General Sir Archibald Wavell should assume the supreme command of American, British and Dutch forces in a theatre of war extending from Burma to the Philippines and as far south as the Timor Sea. This area proved too large and heterogeneous for any system of unified control to be effective, but the principle was afterwards adopted with satisfactory results in other theatres.

The Americans might have been expected to conclude, too, that President Wilson carried neutrality too far when he refused to sanction active preparations for war until well past the eleventh hour. Foreign observers in New York and Washington reported in the spring of 1917 that military unpreparedness could not have been more complete and that officials of the United States Treasury were aghast when they learned that the war was costing the British the equivalent of thirty-five to fifty million dollars a day. But it would be rash to assume that the Allied and Associated Powers, as a whole, would have gained any advantage if American industry had devoted itself between 1914 and 1917 to the task of providing a hypothetical expeditionary force with field, medium and heavy guns, machine-guns, rifles, ammunition, explosives,

* The quotation is from a report submitted to President Wilson by General Tasker H. Bliss, United States Army, in December 1917.

bombers, fighters, scouts and transport. Deprived of some or all of their American-made equipment, the European Allies might have lost the war before the United States had occasion to enter it. Even if that did not happen, the problem of transporting a fully-equipped expeditionary force across the Atlantic might have proved insoluble. A valid lesson which the Americans might have drawn from the experiences of 1914 to 1918—and which some of them did draw twenty years later—was that they must take good care, if history repeated itself, to strike a balance between their own needs and those of potential allies or associates.

America's entry into the war coincided almost exactly with the collapse of Russia. The Russian monarchy was on its last legs when the United States severed diplomatic relations with Germany in February 1917. By the time of the President's announcement to Congress in April, the Tsar had ceased to rule and was virtually the prisoner of a short-lived provisional government. The first American division to reach the Western Front suffered its initial casualties at the very moment when Lenin and Trotsky were preparing to seize power in St Petersburg and Moscow. When Lloyd George told the Americans in November of Britain's shipping problems but urged them none the less to send to France as many men as they could spare, his appeal was prompted by the knowledge that Russia was falling to pieces and was likely to conclude a separate peace with the Central Powers within the next few weeks or months. Little more than a week later, Germany accepted Lenin's offer of an armistice. The Treaty of Brest-Litovsk—which put the Baltic provinces, Finland and Russian Poland more or less at Germany's mercy and promised the Germans access to rich sources of food and raw materials in European Russia—followed in March.

Russia's attempts to equip her armies on a scale suitable for a two-front war against Germany and Austria were hampered from the start by widespread inefficiency and corruption, poor communications, and the difficulty of obtaining delivery in war-time of imported machinery and tools. Since roads were few and mostly bad, Russia depended even more than her neighbours on railways for the carriage of men and goods. Yet her rail network was extremely sparse by German standards. Moreover, Russian trains were notoriously slow. A train carrying the members of a high-level Allied mission from Murmansk to St Petersburg early in 1917 travelled at little more than ten miles an hour. Officials sent to meet it, also travelling by train, arrived twelve hours late.

Goods trains, of course, were slower still. Since enormous distances separated the principal centres of distribution from wartime ports of arrival for cargoes shipped from Britain, France or the United States, weeks or months elapsed before goods wagons despatched from Archangel, Murmansk or Vladivostok were seen again. In these circumstances, far more rolling stock than the Russians possessed would have been needed to maintain a steady flow of traffic.

In the light of innumerable complaints in 1914 and 1915 of shortages of shells, rifles and cartridges, committees were set up in 1916 to organize arms production on a regional basis. This reform achieved only a limited success. At best, local committees could do no more than see that factories under their supervision made the most of what they had. Only a strong, well-organized central government, served by competent officials, could have ensured that the best use was made of a limited carrying capacity to deliver raw materials, fuel and equipment to the factories most in need of them and that available supplies of weapons and ammunition were parcelled out to armies in accordance with a predetermined plan.

No such government existed in the Russia of 1914 to 1917. Nominally the Tsar had ruled since 1905 as a constitutional monarch. In practice, Russia remained a tottering bastion of autocracy until the collapse of the monarchy on 15 March 1917 put power into the hands of a Duma that had learnt to criticize but not to rule. In wartime an autocratic system might not have been fatal to Russian interests if a man of the calibre of Peter the Great had been in charge. But the Tsar Nicholas II was not designed by nature to be a successful despot. Weak, irresolute, easily swayed, but capable of unyielding obstinacy when he believed that his authority was questioned or the future of the dynasty at stake, he had a knack of surrounding himself with unsuitable advisers. After the military setbacks of 1915 he dismissed some officials and made some changes in the High Command, but his reforms were by no means uniformly beneficial. By assuming supreme command of the armed forces and relegating the former Commander-in-Chief, the Grand Duke Nicholas, to a command in the Caucasus he forfeited the confidence of many members of the aristocracy and the middle class. As a military commander the Grand Duke was open to criticism, but he was popular in the army and a staunch upholder of the alliance with France. His relegation to a comparatively minor post gave rise to rumours that the Tsar, prompted by the supposedly pro-German Tsarina and the imperial favourite Rasputin, was contemplating a deal with the Kaiser.

In any case, the Grand Duke was not responsible for the inadequacy of the Russian communications network. And it was this, more than any other single factor, that led to Russia's downfall. In the summer of 1916 the Russians advanced up to fifty miles through the Austrian lines and took 350,000 prisoners before their advance petered out for lack of supplies. A year later they punched a broad gap in the German front, but soon broke off their offensive because they were demoralized by hunger and had ceased to believe that victory could improve their lot or that of the civilians in their rear. On both occasions poor communications were largely responsible for the conditions that precluded a decisive success. If soldiers were poorly fed and civilians cold and hungry, if arms factories closed or went on short time for lack of fuel to stoke their furnaces, the reason was not that there were shortages of food or fuel in the country as a whole. Russia was rich in timber, coal and grain. The loss of her foreign trade made the whole of the wheat previously exported available for home consumption or for storage. St Petersburg and Moscow starved and shivered, troops went on half rations or were given lentils instead of bread, because the machinery of distribution broke down under the stress of war. For the same reason, millions of rounds of ammunition and thousands of tons of explosives shipped to Russia or produced in Russian factories never reached the troops.

The Allies recognized in 1917 that communications were Russia's weak link. They offered to finance the purchase of locomotives and rolling stock. They appointed a mission to reorganize the railways. But these measures came too late. By the time the mission was ready to grapple with its task, Russia was in a state not only of material but also of moral and intellectual chaos. Until the Tsar abdicated, thinking Russians of all shades of political opinion were united in the conviction that he must quit his throne. Once he had left it, there was no unanimity in Russia. Moderates jostled with extremists, Bolsheviks with Mensheviks, professed Socialists who accepted the tenets of the Communist Manifesto with professed Socialists who didn't. Some members of the Duma saw in the collapse of the Tsarist régime an incentive to prosecute the war with renewed vigour. Others argued that, since the imperialist aims with which Russia had entered the war were no longer attainable, the provisional government should aim at bringing the struggle to a close by persuading the belligerents to come to terms. Still others thought the nation should turn its back on the Western Allies and make a separate peace with the Central Powers.

In Western Europe the months immediately preceding the Communist take-over in Russia were marked by developments which cost British and French statesmen some anxious moments. In France there was a mutiny in the army, arising from the failure of an offensive which a new Commander-in-Chief, General Robert Nivelle, had insisted on launching although he had reason to known that strategic surprise had been lost and that few of his subordinate commanders expected decisive results. In Britain a whispering campaign was launched against the coalition government. Lloyd George was alleged to have become the dupe, or even the accomplice, of unscrupulous men who insisted on his fighting the war to a finish because they were interested not in the independence of Belgium, the sanctity of treaties and the liberation of French territory invaded by the Germans, but in 'the oil wells of Mesopotamia'. A Labour member of the War Cabinet, Arthur Henderson, added to the government's anxieties by travelling to Paris, against the express wish of his colleagues, to discuss with French and Russian Socialists arrangements for an international Socialist conference at Stockholm which German as well as Allied and neutral delegates were expected to attend. Arms manufacturers debarred by wartime legislation from making higher divisible profits than they had earned in peacetime were held up to obloquy as callous exploiters of human misery.

The allegation that Lloyd George was willing to prolong the war for the sake of the oil wells of Mesopotamia was, to say the least, a wild exaggeration. In 1917 there were no commercially productive oil wells either in Mesopotamia or in any other part of the Ottoman Empire. The Turkish government had promised a few weeks before the outbreak of war to lease to the British-controlled Turkish Petroleum Company any oil deposits 'discovered or to be discovered' in the administrative districts of Baghdad and Mosul, but many years were to elapse before oil was found there in commercial quantities. British forces had been sent to Mesopotamia in 1914 not to capture oil wells which did not come into existence until more than ten years later, but to safeguard the government-controlled Anglo-Persian Oil Company's assets in Persia and at Basra, and incidentally to encourage Arab chieftains to throw off their allegiance to the Sultan. To these aims had afterwards been added the weakening of the Central Powers by the infliction of a decisive defeat on the Turks. The coalition government was of course aware that Mesopotamia was a possible future source of oil; but the interest taken by Lloyd George's Conservative colleagues in the Ottoman Empire stemmed more from their long-term preoccupation with the defence of India than from any other single factor. Lloyd George himself had always advocated resolute action in the Near and Middle East. His view was that, far from prolonging the war, a shrewd blow struck at Germany's weakest ally would tend to shorten it.

The allegation that the war was being prolonged in the interests of the oil industry was not, therefore, taken very seriously by the coalition government. The poet and infantry officer Siegfried Sassoon, when he claimed that Britain was fighting for the oil wells of Mesopotamia, was not charged with sedition but treated for shell-shock.

Arthur Henderson's flirtation with International Socialism was an embarrassment to the government, but no more than an embarrassment. Henderson was Secretary of the Executive Committee of the Labour Party and a prominent trade unionist. Had participation in the Stockholm Conference become an issue between the government on the one hand and the Labour Party and the trade unions on the other, the outcome might have been industrial strife which could have led to a catastrophic fall in arms production. But that did not happen. Before Henderson made his defiant visit to Paris and allowed himself to be talked over by foreign Socialists, his committee recorded the opinion that British participation in the conference would be unwise. The rank and file of the party came to the opposite conclusion, but voted by only a narrow majority in favour of participation when the matter was raised again after Henderson had resigned from the War Cabinet and was shown to have misled both government and party about the attitude of the Russians. Moreover, many trade unionists, irrespective of their political affiliations, were more than willing to accept the government's view that, if British labour was to be represented at all at Stockholm, it ought not to be represented by a member of the War Cabinet or by anyone empowered to discuss Allied war aims with neutral or German delegates. The Sailors' and Firemen's Union went so far as to give orders to its members that Ramsay MacDonald, spokesman of the small splinter-group Independent Labour Party and notoriously hostile to the coalition government, was not to be allowed to embark in any British vessel for the purpose of travelling to Stockholm, although he held a passport which would have permitted him to make the journey if he had been able to find a ship to carry him.

So Lloyd George's government was not seriously shaken by the assertion that its reasons for wishing to continue the war until the Central Powers admitted defeat were discreditable. The charges brought against arms manufacturers, on the other hand, attracted a good deal of support. At a time when millions of soldiers and civilians in almost every European country were heartily sick of the war and longed to end it, many people were willing to give some measure of assent to the proposition that only arms manufacturers could expect to profit by continuance of the struggle and that therefore only arms manufacturers could wish to prolong it.

The flaw in this argument was the absence of any evidence either that arms manufacturers, as a class, did expect to profit by continuance of the struggle, or that their influence was decisive. After three years of war, probably most of them, and certainly some of them, were just as much tired as anyone else of a conflict which deprived them of access to foreign markets, restricted their earnings while burdening them with more work than wartime staffs could handle without great difficulty, and compelled them to conduct their affairs in accordance not with their own wishes but with those of the statesmen and officials who controlled their allocations of labour and raw materials. Whether the Allied and Associated Powers would have been wise to come to terms with their enemies in 1917 can only be a matter of opinion. Had they done so, a remodelled Austro-Hungarian empire or federal republic might conceivably have become a bastion against German aggression in central and eastern Europe, but Germany would not have been deprived of her economic stranglehold over Belgium. As things turned out, the war continued until 1918 because it was not until 1918 that the German General Staff were prepared to recommend that peace should be made on terms which the Allied and Associated Powers could accept. British, French, and American arms manufacturers concurred or acquiesced in the decision to go on fighting until the Central Powers admitted defeat, but so did British, French and American citizens of all trades and classes. German arms manufacturers shared, perhaps even helped to promote, the belief of the General Staff that hegemony over Belgium was a vital German interest, but they were far from being alone in doing so. The fact that millions of Europeans would have liked the war to end in 1917 did not mean that they were able or willing to put pressure on their governments to end it on terms regarded by diplomatic and military experts as unsatisfactory.

In the outcome, Germany's hopes of settling the Belgian question by gaining a decisive victory on the Western Front before the Americans arrived in strength were frustrated by the stubbornness of British and French troops, and by Lloyd George's equally stubborn insistence on the creation of a Supreme War Council to co-ordinate Allied and American strategy and military policy at the highest level. The Treaty of Brest-Litovsk did not solve the most recalcitrant of Germany's problems as her statesmen expected it to do. The Ukraine, which was to have furnished Germany with vast quantities of wheat, set up a regional government which declared its independence of Moscow, claimed the right to negotiate its own treaties with the Central Powers and ordered farmers to surrender or destroy their crops rather than allow them to fall into German hands. These orders were not always obeyed, but Ukrainian peasants and speculators did not need to be told that hoarding their grain or bartering it for livestock or consumer goods might be more profitable than selling it to the Germans at fixed prices. The Germans did succeed in buying some Ukrainian wheat, but not as much as they needed or expected.

The oil of the Caucasus proved equally elusive. The regional government of Georgia, rejecting the authority of Moscow but professing a loose allegiance to the newly-formed Federal Republic of Transcaucasia, allowed Turkish and German troops to gain a foothold in the Caucasus and signed treaties with both countries, but was thereupon deposed by less pliant leaders, who denounced the treaties and invited a small body of British troops which arrived from Mesopotamia by way of Persia to join the armed forces of the federal republic in turning the Turks and the Germans out. The British responded by repelling a Turkish attack on Baku but afterwards withdrew to Persian territory.

Towards the end of the war the output of the German arms industry fell short of the target set by the military authorities. German troops on the Western Front did not suffer from any immediate scarcity of weapons or ammunition, but would almost certainly have found themselves at a serious disadvantage if the fighting had gone on for another year. The chief cause of the decline was a shortage of labour. This was accentuated by a lack of natural resources which forced manufacturers to rely partly on synthetic materials often inordinately expensive in terms of the man-hours needed to produce them. A contributory factor was the difficulty of providing civilians with adequate supplies of food when meat, grain, sugar and edible fats were scarce as a result of poor harvests, a dearth of fodder, the Allied blockade and the disobliging attitude of the Ukrainians. The armed forces received rations which seemed fairly generous until the capture of supply dumps during the spring offensive in 1918 led to the discovery that British rations were far better. Although this experience did not enhance the confidence of the troops in ultimate victory, they continued to give a good account of themselves until well past the moment in August when the High Command came to the conclusion that Germany could no longer hope to impose her will on the enemy by military means and must aim at a negotiated peace.

Thereafter seven weeks of hard fighting followed before the Germans asked for an armistice, another five before an armistice was granted. The document then signed by representatives of Germany and the Allied and Associated Powers was only one—albeit the most important—of a series of armistice agreements by which all the belligerents undertook to suspend hostilities on the understanding that they were willing to make peace on terms to be discussed, and in Germany's case foreshadowed in long-drawn pre-armistice negotiations.

The terms of the armistice agreements showed clearly that the Allied and Associated Powers regarded themselves as having won the war. Whether they would be able to secure compliance with those terms and go on to make a stable peace was another matter. They had accumulated large reserves of war material for an offensive on the Western Front in 1919, but had nothing like the number of troops they would have needed to occupy the whole of Germany, the Austro-Hungarian and Ottoman Empires and Bulgaria, to say nothing of Russian Poland, Finland and the Baltic states. Nor could they hope to dominate the whole of Europe and the Near East for years to come. Only the governments of the countries concerned could ensure, if they had the will and the means to do so, that orders for the laying down of arms and the disbandment of units were obeyed and that obligations about to be undertaken under the peace treaties were honoured. By the time the plenipotentiaries of the victorious powers assembled in Paris for a preliminary discussion of peace terms there was, however, no government in Vienna capable of shaping events in Bohemia, Moravia, Slovakia, Ruthenia, Galicia, Slovenia, Croatia, Bosnia, Herzegovina or Hungary. In Turkey the Allies controlled Constantinople, both shores of the Dardanelles and the Sultan's government, but were soon to undermine their own and the Sultan's authority elsewhere by sending a Greek army of occupation to Smyrna and thus precipitating a revolution in Anatolia.

EIGHT

Post War

The First World War was brought to a formal close by peace treaties which not even the most rabid opponent of the private manufacture of arms could suppose to have been inspired by an undue regard for the interests of arms manufacturers. Of the three men who played the leading roles in the Preliminary Peace Conference of Paris, it was Lloyd George who most clearly left his mark on the crucial Treaty of Versailles. His policy was to disarm Germany but prevent the French from establishing a dominant position in Europe at Germany's expense. He successfully opposed the creation of French-dominated autonomous republics in the Rhineland and the Palatinate, insisted that Germany's post-war army should be recruited by voluntary enlistment, and joined President Wilson in guaranteeing France against renewed aggression. He recognized that the Germans could not be permanently prevented from rebuilding their armed strength if they had the will to do so, but looked forward to a future in which they would have no incentive to put guns before butter. If the French, reassured by the Anglo-American guarantee, could be persuaded to follow the example of Britain, the United States and the new Germany by renouncing compulsory military service in time of peace, the Germans would not need to rearm. Large standing armies would then become a thing of the past and the demand for land armaments would fall off sharply. In the meantime a naval agreement which Lloyd George hoped to conclude with the United States would substantially reduce expenditure on naval weapons.

This dream was shattered by the predictable refusal of the United States Senate to ratify the Treaty of Versailles and the consequent nullification of the Anglo-American guarantee. The French, deprived of the promise of support for which they had sacrificed their hopes of a strategic frontier on the Rhine, insisted on retaining their large conscript army and for some years declined even to discuss proposals for its limitation. They sought re-assurance in alliances with Belgium, Poland, Rumania and the Austrian succession states, and in 1927 began to build along their frontier with Germany the chain of fortified positions known as the Maginot Line. French industrialists found profitable outlets for their goods or their services as purveyors of technical know-how at home, in French North Africa and Syria and in Eastern Europe, but gladly renewed old associations with German coal and steel magnates when the Locarno agreements of the mid-1920s brought a temporary improvement in Franco-German relations.

Lloyd George was not denied his naval agreement with the United States but was compelled, as part of the bargain, to exchange a long-standing alliance with Japan for the uncertain prospect of American co-operation in the event of a Japanese threat to British interests in the Far East and the South-West Pacific. The Washington treaties imposed severe restrictions on Japanese naval strength, but left Japan in a stronger position to attack American possessions in the Philippines than the Americans were to defend them.

The treaties also imposed severe restrictions on British and American capital ship construction. For Armstrongs and Vickers, this was only one of a series of misfortunes which afflicted British heavy industry between the end of the war and the middle of the 1920s. Before the war, both firms had relied for an important part of their revenue on orders from the Admiralty for warships, guns and gun-mountings. When planning their post-war production and investment programmes both assumed that, irrespective of any further orders they might receive for armaments, the end of hostilities would bring very large demands not only for consumer goods, but also for merchant vessels, locomotives and rolling stock to replace losses arising from enemy action or war-time wear-and-tear. Vickers went to a great deal of trouble and expense to prepare and equip their shipyards and heavy engineering works at Barrow and Sheffield for the production of merchant ships, locomotives, boilers, turbines, reciprocating engines and miscellaneous railway material; their factories in the South of England for the manufacture of light industrial machinery, machine-tools, sewing-machines, domestic appliances, toys, furniture and sporting guns. They also made a massive investment in the electrical industry and the manufacture of rolling stock by acquiring the electrical and carriage-building interests of the Metropolitan Carriage, Wagon and Finance Company, and setting up the Metropolitan-Vickers Electrical Company. Armstrongs were kept busy immediately after the armistice with the building of new ships for the Cunard and Peninsular and Oriental Lines; but they, too, branched out in new directions. In 1919 they entered the motor-car industry by buying the Siddeley-Deasy motor works at Coventry. Under a new chairman, they went on to invest hugely in the development of a paper-mill and ancillary enterprises in Newfoundland.

None of this meant that either Armstrongs or Vickers were prepared to carry such preoccupations to the point where they ceased to be arms manufacturers. Even if they had wished to make such a transformation, they would have had some difficulty in doing so as long as the government of the day had any power to influence their decisions. British statesmen of the early post-war era, reluctant though they were to countenance any form of interference with private enterprise which involved the payment of subsidies or might affect the free play of the market, had no intention of allowing their principal suppliers of arms to go out of business if they could help it. On the assumption that large orders from the British government were likely to be infrequent during the next ten years or more, it seemed to follow that the firms would have to go through the familiar process of maintaining their arms-producing capacity by seeking orders abroad.

Armstrongs and Vickers, it will be remembered, had become joint investors before the war in a number of foreign enterprises. They were, for example, minority shareholders in the Ottoman Dock Company, an undertaking formed in 1914 for the purpose of administering Turkish docks and arsenals. They also held in 1914 one-sixth of the bonds which provided most of the company's working capital. The rest of the shares were held by the Turkish government; the rest of the bonds by private investors. Armstrongs and Vickers, feeling after the outbreak of war with Turkey that many of these people must have been induced to buy bonds that might prove worthless by the knowledge that two highly respectable British companies were associated with the enterprise, offered to buy all bonds which holders cared to sell them at cost price plus accrued interest.[1] The result was that by the end of the war they held more than 90 per cent of the entire issue. No interest having been paid, they became entitled, when the Treaty of Sèvres brought hostilities between Turkey and the Allies to a formal close, to take over the business of the company. To the accompaniment of the Kemalist revolt in Anatolia—precipitated by the decision of Lloyd George, President Wilson and Clemenceau in 1919 to allow Greek troops to disembark at Smyrna—they succeeded for more than eighteen months from the spring of 1920 in running it at a profit. Thereafter conditions in the Near East became steadily more chaotic. By the end of 1921 the French and the Italians were known to be supplying Kemal with arms, although Constantinople was still garrisoned by Allied troops and the Allies still claimed to be upholding the authority of the Sultan. In the following year a confrontation between Kemalist and Allied troops at Chanak put Lloyd George out of office and gave Britain the first Conservative government to rule at Westminister since 1905. An armistice with Kemal and the Sultan's abdication left Armstrongs and Vickers still in control of the Ottoman Dock Company, but soon after the signing of the Treaty of Lausanne in 1923 they paid off the company's overdraft and withdrew from Turkey.

Armstrongs and Vickers were also partners in an undertaking in Japan. This was a company formed in 1907 on the initiative of the Japanese government and known in England as the Japan Steel Works or Japanese Steel Works. Its primary purpose was to cater for the needs of Japan's armed forces and especially those of the Imperial Japanese Navy. The Japanese authorities had sounded a number of American and European companies before calling in Armstrongs and Vickers, dividing half the share capital of £1,000,000 equally between them, and allotting the other half to Japanese investors associated with the Mitsui family. Similar allotments were made when the company's issued capital was increased in 1909 to £1,500,000. During the war, however, the company acquired a number of subsidiary undertakings, and by 1919 its issued capital stood at £4,000,000. A redistribution then allotted shares with a nominal value of £500,000 to each of the British firms. Their combined stake in the enterprise was thus reduced from a half to a quarter.

Meanwhile the company had become sole agent in Japan for both Armstrongs and Vickers. Vickers were much put out when they discovered in 1920 that Armstrongs, without consulting them, had negotiated with the company a new agency agreement from which products other than armaments were excluded, but they felt bound to follow suit.[2] Like the agreement negotiated earlier by Armstrongs, the new agreement they then negotiated with the company was due for renewal in 1922.

All these arrangements were made when Britain and Japan were still linked by an alliance whose origins went back to a pact of mutual assistance signed in 1902. By the time renewal of the agency agreements came up for discussion, the Japanese directors of the company were, however, well aware that the British were under strong pressure from the United States government to abandon the Japanese alliance, and that in all probability it would be allowed to lapse. In any case the naval disarmament proposals then under discussion in Washington would almost certainly have an adverse effect on the company's order book. The company declined, therefore, to renew the agreements. The two British companies thereupon came to the conclusion that they might be well advised to bring their association with the Japanese arms industry to a close, but they were unable to persuade the Japanese to let go of their end of the string. When Vickers suggested in 1923 that part of a large sum which the company hoped to receive from the Japanese government as compensation for cancelled contracts might be used to buy them out, the authorities in Tokyo made it clear that this would not be permitted. For the time being the British companies had no choice but to retain their holding and content themselves with an average annual return of some seven per cent of its nominal value.

Vickers—but not Armstrongs—also had a continuing interest in the Spanish arms industry. This arose partly from their ownership of the Placencia de las Armas Company, partly from the investment they and John Brown and Company had made in 1908 in La Sociedad Española de la Construccion Naval. Originally the two British firms had held about a quarter of the equity, but their share rose by the end of the war to some thirty per cent. The rest of the shares were held by the Spanish government. During and immediately after the war La Sociedad completed for the Spanish Navy the first and second stages of the ambitious rearmament programme which had called it into being. In 1921 the Spanish government embarked on the third stage. Again, large contracts were awarded to La Sociedad. Industrial strife in Catalonia, the Riff insurrection in Morocco and the advent to power of General Primo de Rivera had no immediately adverse effect on La Sociedad's fortunes. For some years Vickers and John Brown and Company drew substantial dividends. They also received the benefit of orders placed by La Sociedad, or directly by the Spanish government, with them or their associated companies. These companies included Placencia, a wholly-owned Vickers subsidiary. At times more of a liability than an asset, Placencia did fairly well from 1910, when La Sociedad became its best customer, until it was hit at the beginning of the 1930s by the backwash of the world economic crisis.

Vickers-Terni—the Italian armament company in which Vickers held a quarter of the share capital and to whose board they appointed three directors out of nine—had a rather chequered career. Although its formation was discussed as early as 1905, the company did not come into existence until the following year. A site at Spezia had then to be bought, and workshops had to be built and equipped. Production of the armament of two Italian warships began in 1910. In 1915 the company began building aircraft with the help of drawings and technical guidance provided by Vickers. But in 1921, after trading successfully for some years, it became insolvent as the result of a decline in the value of its holding in a subsidiary company. A new company, Terni Elettricità, was then formed. Vickers were allotted a block of shares in recognition of a debt owed to them by the old company. Dividends of the order of six to eight per cent were paid between 1922 and 1925.

Canadian Vickers Limited made an equally slow start. The company offered before the war to complete four cruisers and six destroyers within six years for the newly-formed Royal Canadian Navy, but the Canadian government had second thoughts about its naval programme, and the offer was not taken up. On the outbreak of war the only ship under construction was an icebreaker. Between 1915 and 1919 the company built twenty-four submarines, more than two hundred armed patrol vessels and a number of trawlers, drifters and cargo vessels, but no cruisers and no destroyers. It also turned out roughly a million and a half shells or other projectiles. When its naval contracts came to an end in 1919, a start was made with the assembly of non-military aircraft from parts shipped from the United Kingdom. Four years later the company began building its own aircraft and also launched out in other directions, but it proved such a doubtful investment for Vickers that in 1927 they sold it to a group of Canadian businessmen who raised additional capital by issuing bonds on the open market. Nearly thirty years were to elapse before the parent company regained control.

Finally, there was Russia. Vickers were entitled, in return for their help in forming in 1913 a two-and-a-quarter-million-pound company for the purpose of building and equipping three battleships for the Black Sea Fleet, to invest up to three-quarters of a million pounds in the company and to receive a minimum of ten per cent of the profits for ten years. The Communist takeover in 1917 made it improbable that they would ever receive much more than the fee paid to them on the formation of the company. Other foreign investors in Russian enterprises faced an equally bleak prospect. In general, the attitude of the self-appointed leaders of Communist Russia was, to say the least, unhelpful. At a critical stage of the war, they put their country's allies in jeopardy by making a separate peace with Germany. They set up in capitalist countries Communist 'germ cells' whose activities included the dissemination of subversive propaganda and attempts to bribe journalists and trade union leaders and to suborn troops. By disclaiming responsibility for Russia's external obligations, they made it virtually impossible for the British and the French to pay their war debts in full. Hence it is not surprising that, as long as their grip on the whole of the territories over which they claimed dominion remained uncertain, a good deal of support for anti-Communist Finns, Poles, Balts, Caucasians, Ukrainians, Belorussians and Siberians was forthcoming from Allied governments well provided with arms and ammunition left over from the war.

None of this prevented Lloyd George from making strenuous attempts to resume commercial relations with Russia even before the Russian civil war was over. He made some progress, but had great difficulty in persuading Conservative supporters of his government that he was on the right track. Eventually the Soviet authorities became welcome purchasers of British industrial machinery and even of British tanks, but immediately after the armistice Russia was bound to seem to most British industrial and armament firms a lost market.

There remained the possibility that new markets might be found in countries brought into existence by the collapse of the Tsarist and Austro-Hungarian empires, or newly freed from German domination. Not every firm was eager to do business with countries whose political and economic stability was uncertain. But Vickers had never been afraid of a calculated risk. Their outlook was cosmopolitan. They had, in Basil Zaharoff, a foreign adviser whose knowledge of Eastern Europe was extensive and peculiar.

As early as 1919, Vickers learned that the Poles were not content to rely on the French for military aid and would like to set up their own arms factories. In the following year they established contact, through an intermediary, with a Polish industrial firm, the Starachowice Mining Company. Starachowice, after making an independent approach to Schneider, proposed to set up an arms undertaking with financial assistance and expert guidance from both Schneider and Vickers. Agreements were signed and construction of a factory was begun. Starachowice then ran into difficulties, with the result that the project hung fire so far as Vickers were concerned. Later, the Poles placed welcome orders for tanks made by Vickers under patents acquired from the Irish tank-designer Sir John Carden and his partner V. G. Lloyd.

Meanwhile Vickers received from the Rumanian government an invitation to interest themselves in the expansion of Rumania's industrial and arms-making potential. Their response was to invest at first about £50,000, and later more than twice as much, in a Rumanian industrial undertaking, the Aciéries et Domaines de Resita. This was the only steel firm in Rumania, and was understood to stand well with the authorities. According to a letter written by Zaharoff from Monte Carlo, where he was entertaining the Rumanian Prince Stirbey, the Rumanian government intended to rely for the execution of officially-sponsored civil and military engineering projects partly on Resita, but still more on new undertakings to be established with British and French help. Long and complex negotiations followed, but the results were disappointing so far as British participation in the development of Rumanian heavy industry was concerned. This, perhaps, was not surprising in view of the French government's interest in forging strong links with Rumania and its readiness to extend to French firms in search of orders from foreign governments a degree of diplomatic and ministerial support which no British firm in the 1920s could hope to receive from Whitehall.

The invitation from the Rumanian government reached Vickers in 1922. In the same year they received from Belgrade a letter outlining proposals for the establishment of armament undertakings in Jugoslavia. These were to be financed partly by the state and partly by private enterprise. Thereupon they took up shares in the Société Serbe Minière et Industrielle, a Jugoslav subsidiary of Resita whose designation was commonly abbreviated to Sartid. The assumption was that, even if they received no direct orders from the Jugoslavs, they would benefit by the large orders both Resita and Sartid expected to receive when the Jugoslav project got under way. But these hopes were not fulfilled. During the next few years Vickers saw only a negligible return from their investment in Resita. Their investment in Sartid yielded about £800 in the financial year 1923–24, about twice as much in the following year. None of this money came from the sale of arms or from military contracts.

A good deal was made of these Polish, Rumanian and Jugoslav transactions when the propriety of participation by British firms in foreign arms undertakings was called in question some years later. Whether the directors of Vickers Limited acted wisely or unwisely in signing an agreement with Starachowice and investing relatively small sums in Resita and Sartid can only be a matter of opinion. What does seem clear is that, if these transactions were wrong in principle, then the investments made by British firms before the war in armament undertakings in Italy, Japan, Russia, Spain and Turkey must also have been wrong. Yet no one seems to have thought so at the time. And Britain's need of exports, visible and invisible, was certainly no greater before the war than it was between 1919 and 1925.

At home, the two leading British armament firms soon ran into trouble. Vickers were planning in 1919 to build three hundred locomotives a year. After quoting for batches of twenty-five to a hundred they found that, because they had adopted more exacting standards than were customary in locomotive engineering practice, their prices were too high. Armstrongs, who also began by building locomotives to ordnance standards, came up against the same problem. Both firms had great difficulty, after four years of war, in adapting their output to the needs of customers not prepared to pay for the exceptional accuracy, toughness and capacity to withstand rough treatment demanded by the fighting services.

Such difficulties were not confined to locomotives. At the end of the war about two and three-quarter million gross tons of shipping were needed to restore the carrying capacity of the British merchant fleet to its pre-war level. The needs of foreign countries, Germany excluded, were estimated at well over two million tons. Sir James McKechnie, General Manager of the Barrow shipyard and engineering works and a director of Vickers Limited, concluded that the firm was 'practically assured' of ten years' work at Barrow at a profit of not less than ten per cent. This forecast assumed that the carrying trade would prosper and that shipowners at home and abroad would continue to place orders with British firms for roughly two-thirds of all merchant vessels built throughout the world. McKechnie failed to remind himself that Japan and the United States, among other countries, had greatly increased their shipbuilding capacity since 1913. He did not foresee that the post-war boom in international trade would last only eighteen months; that thereafter British shipbuilding firms would be competing with each other and with the rest of the world in a buyer's market; that costs of production would rise sharply as a result of higher wages and increases in the prices of coal and raw materials. Nor did anyone foresee in 1918 that within three years the British government would have a civil war in Ireland on its hands and be facing industrial strife which reached such alarming proportions that troops were ordered home from the Rhineland, Silesia and Malta to deal with a threatened triple strike of miners, railwaymen and dockers.

Armstrongs and Vickers, with their large resources and long record of successful trading, would have been better placed than almost any of their competitors to survive a slump if they had not given hostages to fortune by investing hugely in enterprises outside their usual scope. When Vickers were known to be negotiating for the purchase of the Metropolitan Carriage, Wagon and Finance Company, they were expected to pay perhaps seven or even ten million pounds for the company's 3,657,432 £1 shares. Dudley

Docker, the astute financier and industrialist who controlled Metropolitan, persuaded them to pay just under thirteen million. The transaction was financed partly by the issue of cumulative preference shares which saddled the purchasers with a heavy burden. Despite conditions which ruled out earnings on the scale of the past nine years, Vickers managed in 1920 to make a profit after tax of well over half a million pounds, but nearly three-quarters of this was absorbed by preference dividends. For the first time since 1867, no dividend was paid on the ordinary shares. After paying five per cent less tax in 1921 and 1922 and nothing in either of the next two years, the board invited Dudley Docker and Reginald McKenna, Chairman of the Midland Bank and a former Chancellor of the Exchequer, to join the well-known accountant, Sir William Plender, in looking into the company's affairs and telling them what they should do to put matters right. Mark Webster Jenkinson, an accountant with special knowledge of the arms industry whom they had consulted some years earlier, then made a detailed investigation which strengthened his existing conviction that the company had carried diversification too far for its activities to be adequately controlled by a single board. If the firm remodelled its organization, he told Zaharoff, and if there was some general improvement in trade, then Vickers could again become 'a successful profit-earning undertaking'.[3]

Largely on Webster Jenkinson's advice, the company went on to seek and obtain the permission of the High Court to write two-thirds off the nominal value of its ordinary shares. Most of the shareholders who attended the extraordinary general meeting at which this decision was approved made it clear that they did not blame Douglas Vickers for the company's misfortunes. In point of fact, by the date of the meeting he had ceased to take an active part in the day-to-day conduct of its affairs. In February 1926, after serving on the board for thirty-eight years and as chairman for eight, he retired to the honorary position of President.

He was succeeded as chairman by Sir Herbert Lawrence. A son of Lawrence of the Punjab and trained as a soldier, Lawrence was married to Isobel Mills, eldest daughter of the financial magnate Lord Hillingdon. He had retired from the army before the First World War, entered the world of banking and become in 1907 a partner in Glyn Mills. Rejoining the army on the outbreak of war, he had served at Gallipoli and on the Western Front, and after

Passchendaele had become Haig's Chief of Staff. He was noted for his unswerving loyalty to Haig, who had chosen him on the strength of a friendship which went back to the days when both men were subalterns in the 17th Lancers. He extended the same loyalty to colleagues and subordinates at Vickers. A man who disdained the trappings of power, he had no private office or private secretary at Vickers House, and travelled daily to work by the Inner Circle. But he was not too austere to recognize the value to the company of Basil Zaharoff, for whom he seems to have had a considerable regard. It was Lawrence who sanctioned the expense-allowance of £10,000 a year which Zaharoff received from the beginning of 1927 until his death in 1936.[4]

Armstrongs were far harder hit than Vickers by the post-war slump. Their share capital stood at the time of the armistice at nine million pounds and rose by the end of the following year to more than ten million. They were a firm of much the same stature as Vickers, with much the same capacity. But sound financial control was not their strongest suit. Perhaps it never had been. William Armstrong, neither an engineer nor an industrialist by training, had founded a highly successful engineering and industrial firm without ever quite losing his amateur status. Andrew Noble, who became his right-hand man and succeeded him as chairman, was a brilliant technical innovator with a passion for perfection. Secure in the knowledge that the company was immensely prosperous, he paid little heed to the financial aspects of its affairs. He had the reputation of being a hard taskmaster, but the reason was not that he was mean or grasping. It was simply that he did not share, perhaps was incapable of understanding, the outlook of most of the people who worked for him. To men with nothing but their earnings to support them and their families, the size of the monthly cheque or weekly pay-packet, the number of hours spent away from home, were bound to seem matters of some moment. Such preoccupations were outside Noble's ken. To him the job was everything, the remuneration unimportant. Doubtless one reason was that his own remuneration was always adequate, but probably that reflection never crossed his mind. A zealous autocrat who never spared himself or others, he found it astonishing that there were men on the company's payroll who seemed not to share his view that time not spent at work, or thinking about work, was time wasted.

But if Armstrong was too many-sided, Noble too one-sided, to conform at all points with the orthodox image of the head of a great industrial undertaking, they did have in common an intimate acquaintance with the technical side of the business. Even Stuart Rendel, who regarded Noble as the supplanter of his brother George and disliked him accordingly, did not allege that there was any falling off in the quality of the company's products under Noble's chairmanship. What he did allege against Noble and his henchmen was that they took too much money out of the firm; that they were careless and imprudent in financial matters; that—in marked contrast with the Vickers family—they failed to provide for the future by enlisting coadjutors of the calibre of Sigmund Loewe, Vincent Caillard, Francis Barker and Trevor Dawson. Sir Andrew Noble was reluctant to delegate authority or impart information, kept everything in his own hands, transacted all the company's London business from his club or lodgings 'at the cost of innumerable night journeys'.

There was some justice in these complaints. Noble and his co-directors confessed to financial irregularities when, under pressure from Stuart Rendel, they agreed in 1911 to 'specified introductions of new blood in the executive' on the understanding that no more would be said of large remunerations secretly appropriated in past years. They could not deny that some ventures into which they had poured money had turned out badly. Sir Andrew Noble *was* reluctant to delegate authority, he did try to do too much. As for the alleged failure to recruit new talent, the facts were not in dispute. Recruits enlisted by Armstrong included Stuart Rendel himself and his brothers George and Hamilton Rendel. They also included Andrew Noble. Noble made some good appointments on the technical side, but his principal supporters during the years of which Stuart Rendel complained were his sons Saxton and John Noble, his son-in-law Alfred Cochrane, his sons' tutor John Meade Falkner.

New directors appointed as the sequel to Stuart Rendel's representations included Lord Kitchener's protégé Sir Percy Girouard; a former Secretary of the Committee of Imperial Defence, Sir Charles Ottley; a retired diplomat and Treasury official, Sir George Murray. Doubtless their influence was salutary, but the leopard did not change its spots. When Andrew Noble died in 1915 he was succeeded by Meade Falkner, an

ANDREW CARNEGIE (1835-1919) ish-born American steel magnate manufacturer of armour plate who a fortune on attempts to promote ersal peace

1b PIERRE-SAMUEL DUPONT DE NEMOURS One of the oldest firms in the United States, E. I. du Pont de Nemours and Company is said to have supplied two-fifths of all the explosives used by the Allied and Associated Powers in World War I

ICHARD JORDAN GATLING (1818-) North Carolinan inventor of the ng gun. Armstrong bought the sh rights in 1870

1d SAMUEL COLT (1814-1862) American designer and manufacturer of pistols and carbines. A pioneer of mass production

2a CHARLES-EUGENE SCHNEIDER Became on his father's death in 1898 sole proprietor of the vast industrial empire founded in 1836 by his grandfather and great-uncle. In World War I Schneider et Compagnie was to France what Friedrich Krupp AG was to Germany

2b ALFRIED (ALFRED) KRUPP (1 1887) German steel manufacturer made his first gun about the middl the nineteenth century. He changed name from Alfried to Alfred after a to England in 1839

2c THE KRUPP STAND AT THE PARIS INTERNATIONAL EXHIBITION OF 186 The war of 1866 between Prussia and Austria established Krupp's reputation as an arms manufacturer, although the performance of his guns left much to be desired

.LFRED NOBEL (1836-1896) Swedish
ntor of dynamite and founder of the
el Peace Prize

3b STUART RENDEL (1st BARON RENDEL) (1834-1913) Barrister, statesman and international arms salesman. Became after Lord Armstrong's death the largest single shareholder in Sir W. G. Armstrong Whitworth and Company

3c SIR W. G. ARMSTRONG (1st BARON ARMSTRONG OF CRAGSIDE) (1810-1900) Solicitor and engineer who entered the arms business in consequence of his and his friend James Meadows Rendel's dissatisfaction with the performance of British artillery in the Crimean War

4a COLONEL T. E. VICKERS, C.B. Second-generation steel magnate and chairman of the family firm from 1873 to 1909. His comment on its entry into the arms industry was that shareholders were lucky to have found a means of escaping a serious falling-off in trade at a time of growing competition

4b ZACHARY VASILIOU ZAHAROFF (SIR BASIL ZAHAROFF) (1849-1936) The knighthood and the robes were genuine, but he was a naturalized Frenchman domiciled in France

5a THE SOMME: OCTOBER 1916 An aspect of war not [ea]sily glimpsed from [b]oardrooms or from G.H.Q.

5b SIR HIRAM MAXIM American inventor of Hugenot descent who became a British subject and was knighted. His fully-automatic gun gave war a new dimension

5c SKODA 420-MILLIMETRE (17-inch) HOWITZER One of the 'outsize mortars' borrowed by [t]he German General Staff from the Austrians before World War I to supplement the giant [h]owitzers designed by Krupp to subjugate the Liège forts

6a OLIVER LYTTELTON (1st VISCOUNT CHANDOS OF ALDERSHOT) (1893-1972) British soldier, statesman and of letters. As Minister of Production from 1942, he fac the daunting task of arguing the Americans about allocatic and priorities

6b HARRY HOPKINS Ameri Lend-Lease Administrator Chairman of the Anglo-Ameri Combined Munitions Assi ments Board in World War He is seen here on the steps the United States Embassy London with General Geo C. Marshall, Chief of Staf the United States Ar

7a THE CAMPAIGN OF THE MUD: PASSCHENDAELE, AUGUST 16, 1917
The campaign was launched by British G.H.Q. despite a warning that the weather in northern Flanders broke each year in August with the regularity of the Indian monsoon

7b CASSINO, MARCH 1944
A New Zealand soldier waits while a Sherman tank blasts German snipers hidden in the ruins of the town

8a SAMUEL CUMMINGS Twentieth-century American dealer in surplus and used defence material. He claimed after World War II that his firm could equip a 10,000-man division at an hour's notice

8b MARCEL DASSAULT Immen successful French aircraft manufact whose products include the Mystère the Mirage. He changed his name f Bloch to Dassault after wartime prisonment by the Germans

8c BAC LIGHTNING Supersonic fighter and ground-attack aircraft which was the principal item in Britain's much-discussed package deal with Saudi Arabia in the 1960s

author and bibliophile who wrote scholarly minutes in a longhand derived from the study of old manuscripts. After completing his stint as tutor to Saxton and John Noble, Falkner had become secretary to the company, presumably because he was one of the few men in whom Andrew Noble was willing to confide. He had gone on to join the board. A man of outstanding intellectual gifts, he was at least an adequate chairman during the years when Armstrongs, like Vickers, were virtually under government control. Nevertheless it is hard to believe that his heart was in the work. His interests, apart from palaeography, book-collecting and authorship, included the history of music and of the development of ecclesiastical ritual. When he was not administering the company's affairs or arguing with the government about profits, he lived not merely in a different world but in a world so remote as to be barely discernible.

His successor was a man of a different stamp. Glyn West had left Armstrongs during the war to become Deputy Director-General of Gun Ammunition at the Ministry of Munitions. There he acquired a considerable reputation as one of the 'captains of industry' so unexpectedly admired by Lloyd George. Rewarded with a knighthood, he returned after the war to his old firm. Appointed chairman when Meade Falkner retired in 1920, he followed the example of his predecessors by assuming dictatorial powers. Like Noble he drove himself hard, disliked any form of power-sharing and tended to keep colleagues and subordinates in the dark. There the resemblance ended. Glyn West was not, as Noble was, a gifted mathematician whose mind teemed with ideas for improving the company's products or cutting costs. Nevertheless he was energetic, industrious, not easily put down. In normal times he might have found his feet, have learnt to take advice, have gone on to become a reasonably successful chairman.

But the times were not normal. Because their reputation as builders of ocean liners stood so high, Armstrongs were hit rather later by the post-war slump than some of their competitors. But eventually the slump did hit them. They had then to deal not only with a situation in which their shipyards could obtain further orders only by quoting which yielded scarcely any profit, but also with the consequences of their decision to build and operate a paper-mill in Newfoundland.

This was a gigantic undertaking. The property acquired by

Armstrongs consisted of a large tract of dense forest and the village or hamlet of Cornerbrook. For three or four months out of twelve it was snowbound. A considerable area had to be laboriously cleared of timber before construction of the paper-mill and a hydro-electric power-station could begin. Plant had to be installed under the supervision of experts who could not start work until the buildings were ready. A port with facilities for the handling of heavy or bulky loads had to be established. A whole town, complete with houses, shops, schools and hospitals, had to be built to attract workers who would not come unless they could bring their wives and children. Only careful co-ordination of these activities could avert overlapping which was particularly wasteful in Newfoundland because every labourer, skilled tradesman, engineer or technician brought to the site had not only to be paid, but also to be housed and fed.

Unfortunately careful co-ordination was not forthcoming. Glyn West and his colleagues seriously underestimated the magnitude of the enterprise, the effects of the climate on the rate of progress, the degree of supervision needed to keep muddles, mistakes, confusion and vexation within bounds. No member of the board was qualified by experience or local knowledge to direct, by remote control, so unfamiliar and so vast an undertaking. Unless they were left to settle themselves, questions which called for high-level decisions had therefore to be decided either by directors largely ignorant of conditions in Newfoundland, or by men on the spot imperfectly acquainted with the company's long-term problems and the extent of its resources.

An inevitable consequence was that the cost of developing the site and installing plant and machinery far exceeded expectations. This might not have been disastrous if the enterprise had been financed by speculators willing to risk setbacks in the hope that eventually substantial profits would be earned, as indeed they were when the venture passed into other hands. But that was not the situation the directors had to face in the middle of the 1920s. The greater part of Armstrong Whitworth's share capital was invested in its shipyards and engineering works at Elswick and Openshaw and its other interests in the United Kingdom. To finance the Newfoundland venture, the company raised loans and overdrafts from its bankers and issued debentures which brought its annual liability for debenture interest uncomfortably close to its average divisible profit during the past few years. By the middle of 1925 the company owed the bank £2,600,000. Although the shareholders did not yet know it, in effect control of the company had passed from them to the Governor and Company of the Bank of England.

In these dire circumstances, the bank appointed James Frater Taylor, a specialist in the reorganization of companies which had run into difficulties, to look into the affairs of Armstrong Whitworth and suggest remedies for its ills. The choice was made by Edward Peacock of the bank and approved by the Governor, the redoubtable Montagu Norman. A predictable consequence was that Glyn West relinquished his appointment. He was succeeded by Lord Southborough, formerly of the Board of Trade. Meade Falkner, who had remained a director after his resignation as chairman, also left the board, as did Ottley and Murray. In the light of these changes, Norman decided not to appoint an advisory committee which would almost certainly have recommended a receivership. Frater Taylor joined the company as virtual controller, dividends except those payable on first preference shares were banned, and the bank agreed, with stipulations, to go on financing the Newfoundland venture and to make other advances.

The company's indebtedness to the bank continued, therefore, to increase. Yet its prospects grew no brighter, except insofar as some progress was made at Cornerbrook. On the other hand, it still possessed the means of earning a substantial revenue. The extinction of its warship-building and arms-making capacity would, in the government's opinion, be a national calamity. The question was how the company could be helped to retain its assets and pay the interest on its debentures without plunging still more deeply into debt.

Armstrongs' troubles, Frater Taylor thought, were largely due to their having invested huge sums in undertakings not central to their activities as a shipbuilding, engineering and armaments firm. Much the same could be said of Vickers, except that their difficulties were not nearly as great. A solution which occurred to a number of people interested in these matters was that both firms should rid themselves of enterprises not essential to their well-being and should bring those that were essential into some kind of partnership.

The difficulty was to hit upon terms which could be accepted by both companies and would also be acceptable to the Bank of England. Furthermore, at least the approval if not the active support of the government was deemed essential. Vickers were in so much better shape than Armstrongs that they were not likely to agree to any scheme which gave them less than complete control of any joint undertaking. Their first proposal, submitted by Webster Jenkinson to Frater Taylor in 1926, was that Armstrongs should form a new company to take over their shipbuilding and armament business and that Vickers should acquire all the shares in the new company in return for ordinary shares in Vickers. Webster Jenkinson suggested that the government should be asked to guarantee a stipulated volume of orders for a fixed period, in return for an undertaking by Vickers to maintain the arms-producing capacity of the new company at such a level that output could be rapidly expanded in an emergency. The scheme was turned down by the bank on the ground that there was 'altogether too much Vickers' about it. But the bank could not deny that Vickers were in a position to claim a dominant role. As Frater Taylor himself admitted, they were able to withstand a prolonged siege, and Armstrongs were not.

A less obviously unacceptable proposal made in the following summer was that a new company, with an issued capital of £18,500,000, should be formed to take over not only Armstrong Whitworth's shipyards and engineering and ordnance works on Tyneside and at Openshaw, but also the Vickers works at Barrow, Sheffield, Erith and Dartford. The company was to aim at a minimum annual profit of £1,250,000, which would be attainable only if arms production formed a large part of its business. The government would be asked to pay the company £300,000 a year for five years as the price of its retention of an arms-producing capacity which would otherwise be lost to the nation. In deference to Downing Street's notorious abhorrence of subsidies, the payment would be called not a subsidy but a 'rent'.

Unfortunately this sugaring of the pill proved ineffective. Winston Churchill, Chancellor of the Exchequer in Stanley Baldwin's Conservative government of 1924 to 1929, made it clear that no rent would be forthcoming. This announcement brought negotiations to a standstill. Vickers were not prepared, in the absence of a subsidy or a firm promise of orders from the government, to go ahead with the merger on terms acceptable to the other side. Montagu Norman, observing that the merger was desirable in the interests of Vickers but essential from the point of view of Armstrong Whitworth, the Bank of England and the national interest, then stepped in to save the day. He arranged that, if the new company's profits in any year during the next five years fell below £900,000, the Sun Insurance Company would make up the difference, with the proviso that its contribution in any one year was not to exceed £200,000. The ostensible consideration was an annual premium of £400, but the insurance company appears in fact to have acted as agent and cover for the Bank of England.

That did the trick. On 31 October 1927, Vickers and Armstrong Whitworth agreed to transfer their naval shipbuilding and armament undertakings to a new company, Vickers-Armstrongs Limited. Vickers Limited received five million £1 ordinary and three-and-a-half million £1 preference shares in Vickers-Armstrongs as payment for land, buildings, plant, machinery, tools, patents and goodwill valued at £8,500,000. Armstrong Whitworth, contributing assets valued at £4,500,000, received two-and-a-half million £1 ordinary and two million £1 preference shares which they distributed among a number of companies controlled by Armstrong Whitworth Securities Limited.

In practice the transfer involved much more than naval shipbuilding and armament undertakings. Not even such experienced company doctors as Webster Jenkinson and Frater Taylor could perform a surgical operation which would separate the physical assets and technical know-how required for naval shipbuilding from those required for the building of ocean liners. Nor would it have been practical or economically desirable to split factories, workshops and shipyards along functional lines. The new company took over the whole of the Barrow, Sheffield, Erith and Dartford works of Vickers; the Elswick, Openshaw and shipyard works of Armstrong Whitworth; and certain overseas interests of both companies. These included Placencia, the Vickers holdings in La Sociedad and Terni Elettricità, and the Vickers and Armstrong Whitworth holdings in the Japan Steel Works. Vickers retained their Crayford works and their aviation interests, including their aircraft factory at Weybridge. Among assets retained by Armstrong Whitworth was their locomotive and general engineering works at Scotswood. This had been used during the First

World War as a munitions factory, and was destined to become once more an important centre of arms production.

Assets sold shortly before or after the amalgamation by Armstrong Whitworth or Vickers included the Newfoundland Paper Mill; Canadian Vickers; a controlling interest in Metropolitan-Vickers; a half share in William Beardmore and Company; and the non-ferrous metals business of James Booth and Company, acquired by Vickers in the last year of the war. The business of the Wolseley Tool and Motor Car Company, in difficulties since 1923, was sold to the future Lord Nuffield by a receiver put in by debenture holders.

Even without the assets sold or retained by the contributory companies, Vickers-Armstrongs in the form it then assumed was too large and too heterogeneous an undertaking to be administered with maximum efficiency by a single board of directors. The English Steel Corporation, controlled by Vickers-Armstrongs but with Cammell Laird and Company as a substantial minority shareholder, was therefore created to take charge of Sheffield, Openshaw and the stamping department at Elswick from the beginning of 1929. On the same principle, Vickers joined Cammell Laird in forming the Metropolitan-Cammel Carriage, Wagon and Finance Company to administer and reorganize the rolling-stock interests of both firms. As the result of a decline in the demand for rolling stock in the early 1930s, the company was not very successful, and in 1934 it was wound up. A new company, the Metropolitan-Cammell Carriage and Wagon Company, was then formed to complete the process of rationalization begun in 1928.

By virtue of their holdings in Vickers-Armstrongs and English Steel, Vickers remained after the amalgamation both an important steel firm with a stake in general engineering, and an important armaments and naval shipbuilding firm with a stake in commercial shipbuilding. They also remained, in spite of the upheavals of the past few years, a firm with a rather exceptionally sound financial and administrative structure. Not everyone approved at the time of the thoroughness with which the directors disposed in 1926 and 1927 of assets deemed redundant. As things turned out, these sacrifices put Vickers in a particularly favourable position to survive the simultaneous onset of a slump in commercial shipbuilding, and the reduced demand for naval armaments which followed the signing of the London Naval Treaty of 1930. With more than five million pounds in cash and government securities at their disposal on the eve of the recession, they could afford to accept, or put associated companies in a position to accept, orders which a firm unable to grant extended credit would have had to decline. They never, in any year after the amalgamation, failed to make a profit of at least half a million pounds or to pay a dividend on their ordinary shares. Their average annual profit, after taxation, for the difficult four years from 1931 to 1934 was £556,531, as compared with £912,696 for the previous four. Armaments, including military aircraft, accounted for roughly half the turnover of the parent company and its subsidiary and associated companies.

A strong liquid position had other advantages. One of the most important was that the directors could make plans for the future in the knowledge that they would be able, as soon as they judged that better times were coming, to increase the productive capacity of the constituent companies without having to ask for loans or overdrafts. The freedom of the Vickers representatives on the board of Vickers-Armstrongs to back their judgment was, however, limited to some extent by the necessity of consulting colleagues who were trustees for Armstrong Whitworth debenture holders and therefore inclined to play for safety. The obvious remedy was for Vickers to transform Vickers-Armstrongs into a wholly-owned subsidiary by buying Armstrong Whitworth out. By 1930 Armstrong Whitworth were willing to sell; but Lawrence was determined to secure the best possible terms for his shareholders and refused to be hurried. A good deal of hard bargaining followed before the transaction was completed in 1935. By that time the post-war era was over and the era of rearmament was beginning.

*

The problems Continental arms manufacturers had to face after the war differed considerably from those confronting their British competitors. That was only to be expected in view of marked differences between Britain's economic structure and that of most European countries, and of local variations in the nature and quality of the raw materials available to heavy industry.

The Treaty of Versailles restored to France the territories annexed by Germany after the Franco-Prussian War. This act of reparation was gratifying to all Frenchmen. From the point of view of industrial and economic convenience it had, however, the disadvantage of putting the ironfields of eastern Lorraine on one side of the political frontier and the coalfields of the Ruhr on the other. The French had coalfields of their own in Lorraine, the department of the Nord and Central France, but their production of bituminous coal and anthracite was less than half Germany's. By the terms of the treaty they were granted access to the rich coalfields of the Saar, but on terms which required the inhabitants to decide at the end of fifteen years whether they wished to be French or German or to remain under international rule. The French also received some German coal in part payment of reparations, but this was never a very satisfactory arrangement. German industrialists objected strongly to the scheme, not altogether without reason since everyone knew that it cost far less to send French iron ore to Germany to be smelted than to send coal from Germany to France.

On the other side of the account, the retention by France of her conscript army seemed to promise French arms manufacturers a continuing domestic market for their products. At the same time—and this was even more important—a string of alliances in Eastern Europe assured them of export markets to which officials of the Foreign Ministry and the staffs of embassies would be glad to smooth their path. Schneider, with their risks already widely spread, did not feel the same compulsion as was felt by Armstrongs, Vickers, and to some extent Krupp to launch out in new directions. Le Creusot remained in the post-war era the focus of a prosperous arms-producing and heavy industrial complex whose ramifications extended to almost every department of French industry.

The Locarno agreements, followed by Germany's admission to the League of Nations, presented French industrialists with opportunities of renewing or forming profitable associations with German coal and steel magnates. These they embraced to the accompaniment of a chorus of approval from statesmen on both banks of the Rhine. The time had come, said Aristide Briand, for Frenchmen and Germans to forget their quarrels. Gustav Stresemann declared himself more than ready to become 'the German who makes peace with France'. No one foresaw in 1926 that within ten years Frenchmen linked by commercial ties with industrialists assumed to be contributors to Hitler's party funds would be eyed askance by many of their fellow-citizens.

To German arms manufacturers, their chances of obtaining further orders for weapons and ammunition from the German government in the near future were bound to seem towards the end of 1918 extremely slender. On 8 November—three days before the signing of the armistice agreement—the German government announced the cancellation of all Krupp's contracts and the suspension of payments due for work in progress.[5] With more than a hundred thousand employees on its payroll at Essen alone, the company faced a difficult situation. On the following day the works at Essen were closed for the first time within living memory. Gustav Krupp ruled on that day that all men employed on 1 August 1914 should retain their jobs. Others were to be dismissed with a fortnight's pay and, in appropriate cases, free travel to their homes. A month later he announced an elaborate programme of diversification. In addition to the cutlery for which Essen was famous, the firm's peacetime products were to include agricultural and textile machinery, dredgers, crankshafts, motor-scooters, cash-registers, adding-machines, movie cameras, optical and surgical instruments and lawn-sprinklers.

Gustav is said to have confessed later that he was always determined to preserve the firm's arms-making capacity at all costs. No doubt he was. He is also said to have connived at evasions of the disarmament clauses of the armistice agreement and the peace treaty. This, too, seems highly probable. The Allies were well aware of their inability to enforce a ban on the hoarding of arms. They could, and did, secure the surrender of large numbers of medium, field and heavy guns and military aircraft. They could not prevent the Germans from concealing more portable weapons such as rifles and machine-guns. Nor could they prevent German-controlled firms in countries outside their jurisdiction from manufacturing weapons suitable for the Wehrmacht. Such weapons could not be shipped to Germany as long as the Allies remained vigilant and the German government at least ostensibly law-abiding. But they could be shown to German officers, and German crews could be trained to use them, in countries which were not parties to the Treaty of Versailles and whose governments were not bound to act as watchdogs for Germany's former enemies.

Some governments did try to stop such practices. In the light of complaints that the Swedish firm of Bofors had been controlled for the past eight years by Gustav Krupp, the Swedish parliament passed in 1929 a law forbidding foreign participation in the ownership of arms factories. Gustav is said to have responded to this move by transferring his interest in Bofors to a holding company in whose books no shares were registered in his name.

Attempts by the Allies to prevent the Germans from providing themselves with naval and military aircraft were equally ineffective. By the terms of the Treaty of Versailles Germany was required to disband her air forces and surrender all aeronautical material to the Allied and Associated Powers. She was also forbidden to manufacture or import aircraft, aero-engines or their components, but only for six months. The result was that, although she duly surrendered more than 15,000 aircraft and some 27,000 aero-engines and was in theory debarred from reviving her naval and military air services, at the end of six months there was nothing to prevent her from manufacturing or otherwise acquiring civil aircraft which could be adapted for warlike purposes. In 1922 the Allies tried to remedy this shortcoming by prohibiting the manufacture of aircraft above a certain size, and in 1924 they imposed restrictions on the number of aircraft that could be built and the number of people who could be employed in the aircraft industry.

These arrangements left loopholes of which German industrialists took full advantage. Between 1920 and 1925 Hugo Junkers opened an aircraft factory at Dessau and investigated the possibilities of aircraft manufacture in Sweden; Ernst Heinkel set up factories in Sweden and at Warnemünde; Claude Dornier started production in Italy and Switzerland; Heinrich Focke and Georg Wulf founded Focke-Wulf Flugzeugbau at Bremen; and Willy Messerschmitt took over the Bayerische Flugzeugwerke at Augsburg. By 1926, when a new chapter in Franco-German relations was supposed to have opened and all restrictions on the manufacture of civil aircraft were swept away, Germany already possessed an aircraft industry capable of turning out not only passenger aircraft but also bombers and fighters. She also possessed the nucleus of an Air Staff, formed in 1920 and 1921 within the framework of the Defence Ministry she was allowed to retain.

None of this meant that Gustav Krupp's diversification programme of 1918 was mere window-dressing, as has sometimes been alleged. Just as Vickers included in their list of peacetime products a number of items for which there proved to be little demand in post-war Britain, so Krupp's list included some products that were hard to sell. Fortunately for him, the firm's fortunes did not turn upon the demand for movie cameras and lawn-sprinklers. In the summer of 1919 he received a valuable order for locomotives and goods wagons from one of the firm's oldest customers, the Prussian State Railways. By 1920 he was selling locomotives and agricultural machinery to Russia.

Even so, it would be artless to suppose that spokesmen for Krupp kept nothing back when they gave visitors to Essen the impression that very nearly the whole of the company's resources had been turned over to the manufacture of non-military products. A great deal of the equipment of the ordnance department was scrapped in 1920 under the supervision of the Allied Control Commissioners; but much of this material would soon have been due for the scrap-heap anyway. What mattered was the retention of a nucleus of expertise. The clandestine manufacture of proscribed weapons in Germany itself—as distinct from their manufacture on German account in Sweden, the Netherlands and elsewhere—appears to have begun about 1927. In 1924 a former flying corps officer was appointed head of the Civil Aviation Department of the Ministry of Transport, and in 1926 the government formed the state-subsidized Deutsche Lufthansa and endowed it with a virtual monopoly of commercial air transport. Thereafter military pilots were trained in secret at Lufthansa schools, and a flying training centre for regular officers of the armed forces was established at Lipetsk in Russia. Nearly all the officers who were to hold high positions in the Luftwaffe during the Second World War attended courses at Lipetsk between 1928 and 1931.

Meanwhile conditions in Western Germany were in no sense ideal. During the Kapp Putsch in 1920 the Krupp works at Essen were seized and held for a time by ex-soldiers dissatisfied with their lot. The authorities regained control without much difficulty, but the French insisted on occupying Frankfurt and Darmstadt as a protest against their failure to ask permission before sending troops into demilitarized areas. In the following year, after the

Germans had offered to settle their liability for reparations under the Treaty of Versailles on terms regarded by the Allies as derisory, Duisburg, Düsseldorf and Ruhrort were occupied without incident, but with doubtful legality since Germany had not been declared technically in default.

Soon afterwards the commission to which the Allied and Associated Powers had remitted the question in 1919 suggested that Germany should pay the equivalent of approximately six and a half thousand million pounds. This was about four times as much as had just been offered, but only about half what the Allies had hitherto hoped to be paid. The French then called up 140,000 reservists and announced their intention of occupying the Ruhr if they did not receive satisfaction by 1 May. Lloyd George asked that the Germans should first be given time to accept or reject an ultimatum, and in due course the ultimatum then presented to them was accepted. The mark, which had stood at 20 to the £ before the war and at 224 to the £ at the beginning of 1921, fell by the end of the year to 1,020 to the £.

Lloyd George went on to plan a full-scale international conference at which he hoped to reach a new understanding with the Americans, restore Russia to the family of nations, and settle the problem of reparations and war debts once for all. The conference was duly held at Genoa in the spring of 1922, but proved a fiasco. The Americans did not attend. The French were represented by their Foreign Minister, their new Prime Minister Raymond Poincaré having announced that he considered summit meetings a waste of time. The Russians and the Germans duly came to Genoa, but sneaked off to Rapallo to conclude a pact to which the British, the French and the Italians were not parties.

The Americans, observing that a solution of the reparations problem seemed as far off as ever, then began to press for payment of the debts incurred by the Allies during the war. The British, who were owed more than twice as much as they owed, offered to forego all claims to reparations and war debts in the context of a general settlement, but this proposal was not well received in Washington. As for the French, their response to American hints was to freeze German funds in French banks, expel German nationals from Alsace and Lorraine, and proclaim their intention of providing themselves with 'productive pledges' by invading the Ruhr and establishing a lien on mines, factories and forests. At a meeting called by the British to discuss proposals made by the Germans in the light of advice from independent experts, Poincaré made it clear that he had no faith in such proposals. He then returned to Paris and arranged that the Reparation Commission should declare Germany in default with respect to deliveries of timber due as payment in kind. This presented no difficulty, since the Americans were no longer represented on the commission and the French and Belgian representatives could always outvote their British and Italian colleagues by invoking the French chairman's casting vote. The situation at the end of 1922 was that, apart from deliveries of coal and timber, the Germans had made only one payment of £50,000,000, which they had raised by borrowing in London.

At a further meeting early in 1923, the Conservative government recently installed at Westminster declined to have anything to do with Poincaré's proposals. The French then withdrew to put the finishing touches to their plan, the mark fell to 81,200 to the £, and watchers by the deathbed of the special relationship between France and Britain recorded a *rupture cordiale*.

The French were due to march into the Ruhr on 11 January 1923. On the previous day—the third anniversary of the date when the Treaty of Versailles was deemed to have become effective—the Americans withdrew from the Rhineland their Army of Occupation and their representative on the four-man Rhineland High Commission which administered the occupied zones on behalf of the Allied and Associated Powers. The French and Belgian High Commissioners promptly passed a resolution purporting to weld all the zones into a single Franco-Belgian zone.

French and Belgian troops and officials, accompanied by a few Italian technicians, then moved into the Ruhr to confront six million unarmed inhabitants. Practically the whole of the working population obeyed orders from the German government to down tools and refuse to co-operate with the invaders. The British, describing the invasion as a violation of the rights of Germany's creditors, advised German ministers and officials to resist any attempt to undermine their authority and to aim at a solution which would bring them a moral victory while allowing Poincaré to climb down without too much loss of face. In September the government agreed to call off the general strike and make other concessions. The French and their Belgian allies retained their uneasy hold on the Ruhr, but gained no advantage from it. Nor did they profit from attempts they made to weaken Germany by subsidizing separatist movements in the Rhineland and the Palatinate and by dismissing or deporting officials not amenable to threats or bribes. Poincaré, although discredited by his manifest failure to solve the reparations problem, remained in office until, in 1924, he was swept away by a fifty per cent decline in the value of the franc on the international money market.

The fall of the franc was, however, as nothing compared with the fate of the mark. The German authorities were reported as early as November 1919 to be contemplating a massive depreciation of the currency. In 1923 they financed passive resistance in the Ruhr by the simple expedient of printing paper money with which to pay people to stay away from work. An inevitable consequence was that the mark plunged to the edge of worthlessness. By October it was quoted at the ludicrous figure of 112,000,000,000 to the £. The savings of small investors were wiped out, recipients of fixed incomes were reduced to penury, large numbers of young men who had hoped to enter the professions became candidates for employment in industry or commerce. But the fall of the mark did not ruin the very rich—by definition a class well provided with hedges against inflation in the shape of foreign investments, real estate and possessions readily exchangeable for gold or foreign currencies. Nor did it destroy the fixed assets and technical know-how from which large industrial undertakings derived their wealth.

In 1924 the German authorities put their house in order by reforming the currency on the basis of a new unit of exchange, the Rentenmark. In the same year the Dawes Committee—a body of experts appointed on the initiative of the United States government—proposed that Germany should discharge her liability for reparations by making annual payments on a scale rising from £50,000,000 in 1924 to £125,000,000 in 1929. This proposal was accepted on the understanding that the Germans would not be called upon to pay more in any year than they could find without serious detriment to their economy. The German government then raised large loans in London and New York, used part of the money as backing for the Rentenmark and to meet its obligations under the Dawes Plan, and made substantial grants to industrialists as compensation for loss of profits during the Franco-Belgian occupation of the Ruhr.

In the course of the next few years support from American investors helped to put German industry so firmly on its feet that by 1928 its output was well above the pre-war level, despite the loss of centres of production ceded in 1919. American merchant bankers also busied themselves in the second half of the 1920s with the financing of large Russian purchases of American machinery and raw materials. At the same time, the State Department's policy remained in most respects firmly isolationist. A system of quotas restricted immigration; high tariffs made it difficult for nations indebted to the United States government to discharge their obligations.

Meanwhile the British pursued a policy of rigid deflation which restored the pound to its pre-war value in terms of convertibility to dollars. The return to the gold standard was not the result of a sudden brainwave on the part of the Chancellor of the Exchequer and the Governor of the Bank of England, as is widely supposed even today. It was the fruit of ideas which had been simmering in the minds of Treasury officials ever since the armistice. It was intended to reaffirm and consolidate the position of the City of London as the world's most important financial centre, but it had the effect of making British exports too expensive to be competitive. Only orders from the Admiralty for cruisers, destroyers and submarines enabled Vickers-Armstrongs to keep their Barrow shipyard open in the late 1920s.[6] By quoting prices which barely covered costs and overheads they did succeed in obtaining some orders for liners and other merchant vessels, but they were forced to close their naval yard on Tyneside from the spring of 1928 until the spring of 1930. Most British shipbuilding firms fared still worse.

Towards the end of 1927 Montagu Norman and his German counterpart, Hjalmar Schacht, crossed the Atlantic to discuss their problems with the arbiters of American monetary policy. The American authorities agreed to take steps whose most noticeable effect was a sharp rise in the prices of domestic securities. The boom continued almost without check for the better part of two years, although by the spring of 1929 officials of the Federal Reserve Board were alive to the possibility that the slump which must follow any further inflation of the stock market might have disastrous effects on the economy. Despite warnings against the dangers of uncontrolled speculation, banks throughout the United States lent money freely to customers who gambled on the hope of further rises in the prices of stocks already overvalued. Some investors began as early as 1928 to curtail or recall advances to German industry in order to take advantage of opportunities at home.

The slump tentatively predicted in the spring of 1929 came in October. On 24 October banks and finance houses, judging that the investment boom had reached its peak, began calling in loans repayable on demand or at short notice. A desperate scramble to unload holdings whose value was tumbling almost hourly came to a climax on 29 October, when more than twice as much business was done on Wall Street as on any previous day. By the end of the month the *average* value of securities quoted on the New York Stock Exchange was down by forty per cent. Many thousands of investors lost everything they had. Jobbers unable to renew loans on which they relied to finance their day-to-day transactions were forced out of business. Some banks suspended payment as a result of the inability of customers to meet their obligations and of panic withdrawals by depositors who still had funds to draw upon.

Not surprisingly, the failure of the federal authorities to avert the crash or mitigate its effects attracted a great deal of unfavourable comment. Some critics of the establishment even accused the Federal Reserve Board of conspiring with financiers at home and abroad to precipitate a crisis. Orthodox economists did not admit that anyone had acted improperly, but confessed that starting a boom was an easier and less painful process than stopping one.

The ultimate consequences of the crash were far-reaching and calamitous. The world prices of most raw materials fell between 1929 and 1933 by some fifty to sixty per cent. Mines, foundries, workshops, factories and shipyards closed, or their output was drastically reduced. Farmers and dealers in agricultural produce were ruined by a spectacular fall in the price of wheat. Manufacturers of consumer goods were left with huge stocks of merchandise which their customers could not afford to buy. By 1933 fifteen million people were out of work in the United States alone, at least thirty million in the capitalist world as a whole.

NINE

Prelude to Rearmament

One consequence of the world-wide distress that followed the Wall Street crash of 1929 was that pressure was put on the governments of capitalist countries to create jobs by spending public money freely on projects that private enterprise could not afford or was not authorized to undertake. Such expenditure was bound to be regarded in some quarters as inflationary or even socialistic; but the only alternative, unless the unemployed and their families were to be left to starve, seemed to be to condemn millions of people to subsist for an indefinite period on some form of public relief, in the hope that eventually trade might improve without government intervention.

Rearmament was the classic method of priming the pump in a country whose economy had grown sluggish. An advantage of this remedy was that, although expenditure on armaments was just as inflationary in tendency as any other form of public expenditure, a government wedded to capitalist principles could undertake it without rendering itself obnoxious to right-wing supporters by committing the state to direct competition with private enterprise. Statesmen who tried to boost trade by placing orders for guns and warships with armament manufacturers and shipbuilding firms might be guilty of all kinds of improprieties, but could scarcely be accused of toying with 'state socialism'.

But the launching of a major programme of rearmament would have been quite out of the question for any American government in the early 1930s. Within the last few years the United States had taken the lead in persuading sixty nations to sign, with tacit or explicit reservations, a document renouncing war as an instrument of policy. The Chiefs of Staff had plans in their files for war with Japan, and even for war with Britain; but no American statesman of the day had any thought of precipitating a conflict with either country.[1] War and preparations for war were out of fashion. The nation's naval strength was below the limits permitted by treaty; standing armies were still considered dangerous to liberty. The Republican President Herbert Hoover saw in the latter part of 1932 that in the prevailing climate of opinion he could meet demands for a lead from the federal government only by initiating expenditure on public works of a non-military character. He was, however, soon put out of office by the swing of the political pendulum. It fell to his successor, Franklin D. Roosevelt, to implement the policy of federal intervention with a thoroughness which appalled most Republicans and shocked even some members of his own party. Roosevelt succeeded in mitigating the effects of unemployment, but not in curing it. Six years after he introduced his New Deal there were still nearly eleven million Americans out of work.

At the beginning of the 1930s the British were equally reluctant to rearm, but for rather different reasons. With a declining export trade, more than two million people out of work, far-reaching responsibilities for the defence of an overseas empire and an obligation to go to the aid of France or Germany should either be attacked without provocation by the other, they might have been expected to jump at an opportunity of helping industry by refurbishing their depleted armoury while prices were still low. But circumstances did not favour such a solution of their problems. A full-scale disarmament conference, at which they hoped to persuade the French to reduce their land forces, was due to start in 1932. In the meantime the publication of officially-sponsored reports on the country's economic and financial difficulties led to speculation against sterling and a sharp fall in the gold reserves. Ramsay MacDonald, the Socialist leader of a minority government, was at heart a pacifist. In any case he was well aware that his government would not last long if it embarked on any form of inflationary expenditure while the pound was under attack. When the government did fall in the late summer of 1931 because its members could not agree on cuts in unemployment benefit, the King asked MacDonald to form a National government pledged to keep sterling linked to gold at $4·80 to the £.

This the new administration failed to do. On 21 September Britain went off the gold standard. The result was that, although her industrialists had to pay more for imported raw materials, they were able for the first time for some years to quote highly competitive prices for exports. A consequent slow improvement in trade antedated by at least a year or two the beginnings of rearmament.

In Germany, the government negotiated shortly before the Wall Street crash an agreement by which the Allies drastically reduced their claim to reparations and undertook to withdraw their armies of occupation ahead of schedule. A scheme which provided for the payment of annual instalments over a period of nearly sixty years was accepted by the Reichstag, although the National and National Socialist parties, led respectively by Alfred Hugenberg and Adolf Hitler, temporarily joined forces to oppose it.

As things turned out, the scheme had soon to be shelved. By 1931 a quarter of Germany's workforce was unemployed and the Chancellor, Rudolf Brüning, was struggling to ward off a threatened collapse of the mark by increasing taxation and reducing unemployment benefit. The French were willing to help with a loan in return for political concessions, but their offer was rejected when Hitler and his associates made it clear that they would not countenance such a deal and would repudiate it if they came to power.

In the meantime clashes between Communist gangs and Hitler's private army, the SA, were a constant source of anxiety to the government. Brüning, reflecting that nearly six and a half million electors had voted for the National Socialist Party in 1930, hesitated to take repressive measures against the SA, but his hand was forced when the Prussian state government produced evidence of its subversive tendencies and threatened independent action.

In May 1932, the aged President Hindenburg forced Brüning to resign by refusing his assent to further increases in taxation and cuts in unemployment benefit. Franz von Papen, a former diplomat with little political following but considerable influence in right-wing circles, then took office with the intention of relying on the President's special powers to keep him in the saddle until he could put through constitutional and electoral reforms which would make his position more secure. His plan for dealing with the National Socialists was to starve them out. Armed by Hindenburg with authority to dissolve the Reichstag, he succeeded in depleting their funds by forcing them to fight two general elections in seven months. They were known to have received substantial contributions from the millionaire industrialist Fritz Thyssen and other sympathizers, but the second of these elections, held in November, left their finances in poor shape. Moreover, as champions of the Right against Communism they seemed to have lost ground to the National Party. Their representation in the Reichstag had fallen since May from 230 to 196 seats, the number of votes cast for their candidates from 37 to 33 per cent of the total.

The fact remained that Papen was still in no position to govern without recourse to presidential decrees. Major-General Kurt von Schleicher, a politically-minded soldier who had become a power behind the scenes by acting as intermediary between the armed forces and the government, argued that this put Papen in an untenable position. Papen agreed after the November election to stand aside while the President made one more attempt to find a Chancellor capable of forming a majority government, but was confident that the attempt would fail and that he would soon be back in office. When it did fail and the President proposed to recall Papen, Schleicher protested that a further spell of government by decree would be an abuse of powers vested in the President to enable him to tide over an emergency.

In May Schleicher had persuaded Hindenburg to get rid of Brüning by telling him that Brüning no longer enjoyed the confidence of the army. He now used the same weapon against Papen, adding that in any case rule by decree had become unnecessary, since he himself was quite capable of forming a majority government. He then invited Gregor Strasser, leader of the Socialist wing of the National Socialist Party, to join him as Vice-Chancellor in a government which would seek support from all members of the Reichstag except those on the extreme Left and the extreme Right. According to a statement attributed to Thyssen, the invitation to Strasser was issued on the advice of Gustav Krupp.[2]

Unfortunately for Schleicher's hopes, Hitler and nearly all his followers responded to this move by disavowing Strasser. On 28 January 1933, after the Reichstag had given a cool reception to the government's proposals for wage stabilization and agrarian reform, the President forced Schleicher to resign by refusing him a dissolution.

Schleicher's departure was not, as might have been expected, a pre-arranged prelude to Papen's recall. Since November Papen had come to the conclusion that he could not afford to risk splitting the country from top to bottom by undertaking a further term of office as leader of a government supported by only a small minority of the Reichstag. The only acceptable way of preventing the Nazis from making life intolerable for any administration not of their own choosing, he now felt, would be to bring them into a government dominated by nominees of the President and the National Party. Since Hitler would not accept any office save that of Chancellor, Chancellor he would have to be.

As early as 4 January Papen had a long talk with Hitler at the house of the merchant banker Kurt von Schröder. Soon afterwards, the National Socialist Party became mysteriously in funds again. The terms of an agreement between the National and National Socialist parties were negotiated later in the month at meetings attended by members of the President's intimate circle. Finally, on the day of Schleicher's resignation, Hindenburg and Papen agreed that Hitler should be installed in office as Chancellor and leader of a Nationalist-National Socialist coalition.

What part, if any, Gustav Krupp played in the transaction which restored the National Socialist Party's fortunes at the beginning of 1933 is not clear. If he was indeed chief architect of the brief alliance between Schleicher and Strasser, one might expect him, as long as the alliance lasted, to be chary of giving financial aid to a party whose cohesion it might destroy by driving a wedge between Hitler and Strasser. At a conference between the Führer and a number of leading industrialists some six or seven weeks later, Gustav thanked Hitler for giving him and his colleagues a clear picture of 'the conception of his ideas' and (according to Schacht) promised him their support; but by that time the alliance was over and Hitler was Chancellor.

What does seem clear is that accounts of the events of 28 January which depict landowners and industrialists as urging the President to put Hitler in power because he was 'the one man able to save the Fatherland' must be received with caution. No one knows what unnamed territorial and industrial magnates may have said to Hindenburg, or what effect their remarks may have had upon him. In the absence of such evidence, it would seem unreasonable to deny that Hitler was made Chancellor for the obvious reason that the President and his advisers could see no other way of carrying on the government of the Reich without giving the Nazis a pretext for making themselves even more obnoxious than they had already shown themselves to be. If Hindenburg and Papen had really wished to make Hitler all-powerful they would scarcely have insisted, as they did, that only three of eleven ministerial posts in the new government should go to National Socialists. Nor would they, one may think, have summoned such an experienced public servant as Konstantin von Neurath from Geneva to pass judgement on the deal, or have stipulated that he and General Werner von Blomberg—both staunch Conservatives—should be respectively Foreign Minister and Minister of Defence.

When Hitler accepted these conditions and agreed that Papen should be Vice-Chancellor and Minister-President of Prussia, with the right to attend interviews between the President and the Chancellor, Hindenburg and Papen had some reason to suppose that they had lured him into a trap from which he would not escape without great difficulty. However, if they did hope to tame him by giving him the semblance but not the reality of power—and Hindenburg, at least, can scarcely have had any other motive—then they made two serious mistakes. The first was that they allowed Hitler's henchman, Herman Göring, to become Minister of the Interior in the Prussian state government—a post which enabled him to dismiss officials hostile to National Socialism and make other significant changes. Their second mistake was that they under-estimated the cunning and mendacity of the man they had to deal with. Since the new government did not command a majority in the Reichstag, the Cabinet agreed at its first meeting that Hitler should try to win the support of the Centre Party. If he failed, a general election would be held, but irrespective of its results the composition of the Cabinet would remain unchanged. By falsely declaring that negotiations with the Centre Party had broken down when in fact they had only just begun, Hitler tricked Hindenburg into announcing, before a denial could be issued, that the Reichstag was to be dissolved. He thus put himself in a position to fight an election with the resources of the state at his disposal.

Also, Hitler was lucky. Shortly before polling-day, and only three days after the police had raided the headquarters of the Communist Party in Berlin, a half-crazed extremist who disapproved of the alliance between Hitler and Papen set fire to the building in which the Reichstag held its sessions. This was such a fortunate stroke for Hitler that he and Göring were suspected of adding complicity in arson to their crimes. Profiting by the almost universal assumption that the Communists had staged the fire as a reprisal for the raid on their headquarters, Hitler persuaded the President to sign decrees restricting personal liberty, the freedom of the press, and the rights of association and assembly promised in the constitution.

Even so, the election did not bring the National Socialists the sweeping victory predicted in some quarters. Gustav Krupp is said to have promised a contribution of a million marks to Hitler's electoral fund, presumably as a donation from the *Reichsverbund*, or Federation of German Industries, of which he was chairman. Göring prepared for polling-day by recruiting fifty thousand special constables, of whom four-fifths were drawn from National Socialist para-military formations. Many complaints were made of attempts to intimidate left-wing candidates and their supporters. Yet the National Socialist Party won only 288 out of 647 seats in the Reichstag.

But Hindenburg's decrees and Göring's stranglehold on the police put Hitler in such a strong position to stifle opposition that the figures hardly mattered. Undeterred by the knowledge that fewer than 44 per cent of the electors who managed to reach the polling stations had voted for his party, he asked for dictatorial powers for four years. By the time the Reichstag met on 23 March to debate the matter, nearly a hundred deputies had been arrested or gone into hiding. An assembly cowed by threats of murder and arson chanted by storm troopers detailed to guard the building passed the measure by 441 votes to 94. Hindenburg gave his assent on the strength of a promise that nothing unorthodox would be done without consultation.

Thus armed, Hitler proceeded to transfer to the central government the powers hitherto vested in the state governments; suppress all political associations except his own party; weld the trade unions into a comprehensive Labour Front; and win Blomberg to his side by promising to disband the SA. This seemed to Blomberg and other senior officers a valuable concession, because it relieved them of the fear that sooner or later civil strife arising from a clash between the SA and the regular army might undermine their capacity to defend the frontiers of the Reich.

In the summer of 1934 Hitler redeemed his promise to Blomberg by instigating the assassination of the leader of the SA and other persons with whom he had scores to settle. The SS, under Heinrich Himmler, continued to flourish but was not regarded in 1934 as tarred with the same brush.

What Gustav thought of these events can only be surmised. In the spring of 1933 he had acquiesced in the expulsion of Jews from the *Reichsverbund*, which then became the *Reichsgruppe Industrie*, with Gustav as its leader. His eldest son and heir apparent, Alfried, had joined the SS while studying engineering at a technical college in 1931. Inasmuch as they were arms manufacturers, the Krupps could not fail to be interested in Hitler's plans to free the Fatherland from the shackles of Versailles. But the clandestine rearmament of the Reich was not, of course, initiated by Hitler, and no one in Germany thought of him in 1933 as an advocate of war with the Western Powers. In his book *Mein Kampf* he argued that the Kaiser and his ministers ought never to have committed the country to a two-front war by making an enemy of Britain. The course he seemed to favour when he came to power was that Germany should gain additional *Lebensraum* by expanding eastwards, if not with the blessing of the British and the French, at any rate with their tacit consent. Germany had undertaken at Locarno to seek revision of her eastern frontier only by means short of war, but her neighbours in the east were allies of the French. It was hard to see how she could expand in that direction without embroiling herself with France, then still the strongest military power.

When the Disarmament Conference opened at Geneva on 2 February 1932, Brüning was still Chancellor of the Reich, but the large number of votes cast at the last election for extremist candidates of the Left and Right suggested that his days were numbered. The British were aware of numerous infractions by Germany of the disarmament clauses of the Treaty of Versailles. Nevertheless they still felt that it was the insistence of the French on maintaining a large conscript army, rather than German revisionism, which 'kept all Europe in a state of tension'. They aimed, therefore, at persuading the French to reduce the strength of their land forces and to concede to Germany the right to rearm on a scale to be agreed. If agreement could be reached on those points, the way might be clear for a general settlement of Eastern European problems on the lines of the Locarno pacts.

In the light of hindsight, it may seem that the Disarmament Conference was bound to fail. But that did not seem obvious to everyone at the time. Both the French and the Germans were willing to negotiate, although both made stipulations. The French insisted that Germany must demonstrate her good faith before France parted with a single gun or soldier. The Germans, arguing that discussion would be pointless unless the inadequacy of the forces they were allowed by the terms of the Treaty of Versailles could be assumed, suggested that the first step should be to form a just estimate of their needs. Attempts to arrive at a mutually acceptable figure by negotiation led to such lengthy discussion of points of detail that the Germans accused the French of obstructionism and walked out of the Conference. With some difficulty, the British persuaded them to return by assuring them that their former enemies regarded them as entitled to equal rights in a system which would provide security for all nations.

The kind of system the French had in mind was one in which the League of Nations would control an international peace-keeping force. The British felt that any such system must involve too radical a surrender of sovereignty to be acceptable, and in any case would be unworkable, especially as the United States did not belong to the League. In the spring of 1933 they put their cards on the table by proposing that France and Germany should each be allowed 200,000 troops for home defence and the French an additional 200,000 to look after their colonial possessions. But by that time Brüning was out of office, Papen and Schleicher had come and gone, and Hitler was Chancellor. In October he took Germany out of the Conference and announced her impending withdrawal from the League of Nations.

In the light of these developments and of admissions by German officers that Germany was already building military aircraft, erecting fortifications and stepping up the training of recruits, the British appointed in November a Defence Requirements Committee whose task was to say what should be done to repair the worst deficiencies in the national defences. But this did not mean that either the British or the French were prepared in 1933 or the early part of 1934 to commit themselves to substantial programmes of rearmament. The French believed that their existing armaments would continue to give them an adequate margin of superiority over the Germans for some years to come. The British still hoped to negotiate an arms convention to which the leading European powers would subscribe. Even a convention which made substantial concessions to Germany would, they felt, be preferable to none. If it did nothing else, at least it would show that an honest attempt had been made to redeem the promise of multilateral disarmament implicit in the Treaty of Versailles.

But they failed to bridge the gulf between French fears and German aspirations. Hitler offered early in 1934 to enter into a ten-year convention on terms which would not require the French to sacrifice any of their heavy armament before the end of the first five years. The French government, allegedly influenced by industrial magnates with a vested interest in the arms trade, refused to make such a bargain at the cost of condoning past infractions of the Treaty of Versailles. The Germans then made it clear that, unless the British could persuade the French to change their minds, they would rearm at their own pace.

On the far side of the Atlantic, the failure of the Disarmament Conference engendered a revulsion from European entanglements which was damaging to President Roosevelt's hopes of leading his people out of the wilderness of isolation in which they had dwelt since the Senate repudiated the Treaty of Versailles. The United States government approved of Britain's attempts to promote European concord. In the spring of 1933 Roosevelt went so far as to suggest that the United States might, in certain circumstances, abandon strict neutrality to help a victim of aggression.[3] But American observers were dismayed by the obvious reluctance of the French to make concessions to Germany. They were shaken by Germany's withdrawal from the conference and the League of Nations, shocked by the Stavisky scandal in France and its aftermath of violence. Not surprisingly, they recoiled from the prospect of involvement in European squabbles. A sign of the times was that the Senate decided in April 1934—a few days before the French finally rejected Hitler's offer of a two-stage convention—to appoint a seven-man committee to investigate the American arms trade and examine the case for a government monopoly of arms production.

This was the United States Senate Munitions Committee, generally known as the Nye Committee because Senator Gerald P. Nye, of North Dakota, was its most active member. Its Chairman was Senator Arthur H. Vandenberg of Michigan. The committee's Assistant Legal Adviser was Alger Hiss, many years later convicted on doubtful evidence of perjury because he denied having passed classified information to the one-time Communist agent Whittaker Chambers.

The committee interrogated not only representatives of armament firms but also federal and local officials.[4] It examined, among other material, thousands of documents afterwards published in some thirty volumes. Much of the evidence tendered to it was of little more than domestic interest, but some threw a startling light on the economics of the arms trade, and some attracted world-wide attention.

Representatives of E. I. du Pont de Nemours, for example, testified to the staggering profits the company had made between 1915 and 1918 by supplying about forty per cent of all the explosives used by the Allied and Associated Powers during the First World War. During those years the company had paid dividends equivalent to some four-and-a-half times the par value of its original capital and had also accumulated large reserves. They mentioned links with Imperial Chemical Industries and a German dynamite company. They admitted contracting in 1933 to sell arms to Germany, but said that they had stipulated after the original contract was signed that no sale should be made without the consent of the United States government, and after consulting Imperial Chemical Industries had cancelled the deal and compensated their local agent.

Evidence about sales of aircraft, aero-engines and components to foreign governments or firms was given on behalf of a number of manufacturing companies. Representatives of the United Aircraft and Transport Corporation and two subsidiary companies, the Pratt and Whitney Aircraft Company and the United Aircraft Company, estimated that thirty-six per cent of all aeronautical equipment produced in the United States in 1933 had been exported and that eighty per cent of the military aircraft bought by the Chinese in that year were of American manufacture. Companies of the United Aircraft group had shipped hundreds of aero-engines to Holland and Germany since 1926, and the Germans were still the group's best European customers; but all shipments had been approved by one or more of the service departments. Spokesmen for the Curtiss-Wright Export Corporation testified that in 1933 or thereabouts five or six 'American aviation people' who were not officers on the active lists of the army or the navy had gone on the pay-roll of the Chinese government for the purpose of organizing a Chinese air force. The United States government had helped the corporation by endorsing its products, by allowing officers on leave to make demonstration flights, and sometimes by lending spares.

Other statements made to the committee pointed to a tendency on the part of American arms manufacturers and their agents abroad to believe that European competitors were not above combining to squeeze them out, received more help from their governments than American firms could count on receiving from the government of the United States, and were addicted to various unsavoury practices. An agent of E. I. du Pont de Nemours and Company—apparently referring to the attempts made by Vickers after the war to gain a foothold in Rumania and Jugoslavia—alleged that Vickers did business in the Balkans by sending Zaharoff to bribe officials on a vast scale. Representatives of the American Armaments Corporation were convinced that a mysterious tie between Skoda and Vickers accounted for their success in the South American market. Two agents of Federal Laboratories Incorporated expressed their belief that the French government acted as broker or selling agent for Schneider. Louis L. Driggs, President of the Driggs Ordnance and Engineering Company, was asked about a letter written from Turkey in 1929 by a local representative of the company. In this 'the Vickers crowd' were accused of making liberal use of 'women of doubtful character' to gain their ends and were said to have the equivalent of an entire embassy at their disposal. Since there was reason to suspect the writer of seeking an excuse for his failure to clinch a contract, probably most members of the committee did not give much more credence to these charges than they did to the wild assertion that King George V was a Vickers agent.

The committee also had before it letters exchanged between

Vickers and the Electric Boat Company. In one of these Lawrence Y. Spear, a vice-president of the American company, had written that 'the real foundation of all South American business' was graft. In his evidence he said that the company paid commissions on South American business, but that he did not call that bribery. He and his co-directors gave a remarkably full account of the company's revenues during the past fifteen years from contracts to build submarines, royalties on submarines built under licence in various countries, and its share of the profits on submarines built by Vickers. But they did not make it clear—indeed, Spear seems not to have been aware—that the company's arrangement with Vickers had been made with the blessing of the Admiralty and that the British authorities had always known that Vickers paid the company a share of the profit on each Barrow-built submarine in lieu of a royalty. The result was that thousands of readers of newspaper reports of the proceedings received the impression that Vickers and the Electric Boat Company were linked by a clandestine agreement whose effect was to increase the cost of their products to their respective governments.

Letters from Sir Charles Craven of Vickers to Spear, which were read into the record of the proceedings, did nothing to correct that impression. Written over a period of some years, they were informal but by no means artless. Their ostensible aim was to keep Spear informed of the tactics used by Vickers to sell submarines, but their real purpose was to convince him that sums payable by Vickers under the agreement ought to be reduced.[5] In some of them Craven had gone out of his way to stress the magnitude of the effort needed to secure orders when business was slack and the excellence of his and his firm's relations with the Admiralty and other customers. This was unfortunate, because the truth was that Vickers owed their success as tenderers for the building of submarines not to personal relationships but to the indisputable fact that submarines could be built more efficiently and more cheaply at Barrow than anywhere else.

Moreover, some of the letters contained passages which were not merely open to misconstruction but seemed positively to invite it. Members of the public who did not know, or had forgotten, that in 1927 Vickers and Armstrongs were in the process of amalgamation were shocked by what seemed a flagrant example of price-fixing when they learned that in that year

Craven had written that Armstrongs would quote 'whatever price I tell them' when responding to an invitation from the Admiralty to tender for submarines. Similarly, when they read that Craven had written in 1928 that he was trying to 'ginger up the Chileans to take three more boats', they concluded that he had been indulging in high-pressure salesmanship. In fact, he had merely urged the Chileans to pay for three submarines delivered to them, so that the three needed to complete their order for six could be completed and delivered.

More damaging were passages which seemed to imply that Craven was or had been in a position to exert backstairs influence at the Admiralty. 'I wonder whether you have heard,' he had written to Spear in 1928, 'that our old friend Percy Addison is now the Director of Dockyards. I helped him all I could to get the job.' Since he did not add that he had done nothing more sinister than help Addison to qualify himself for the post by coaching him in modern business methods, the implication seemed to be that he had canvassed the authorities on behalf of a candidate from whom, presumably, he expected reciprocal benefits. In one letter there was even a passage which seemed to hint at a successful attempt to bribe serving officers or civil servants. Writing to Spear about an order received in 1930, Craven had added to his customary plea for a reduced payment the explanation that 'certain action which involved expenditure' had been necessary. When Spear was asked by the Nye Committee what this meant if it did not mean that Craven had bribed the Admiralty, he said that he did not know. Anyone familiar with the system of awarding contracts at the Admiralty could have testified that it made bribery virtually impossible, but no such witness was called.

The Nye Committee's investigation did not lead to any significant change in the structure of the American arms industry, but was followed by legislation designed to prevent the government from embroiling the country in a European war. In the meantime the Craven-Spear correspondence had a considerable influence on people in Britain who were already inclined to take an unfavourable view of the private manufacture of armaments. The effect was enhanced by the more or less simultaneous appearance of a number of books and pamphlets in which arms manufacturers were depicted as heartless and unprincipled money-grubbers who would stop at nothing to increase their profits. The authors of some of these publications were able and sincere, but few of them adduced any evidence to support their charges, unless the repetition of well-worn anecdotes about the wiles of arms salesmen could be called evidence. For most of them it was enough that the League of Nations had described the private manufacture of arms as open to grave objections and that the Temporary Mixed Commission on Armaments had listed a number of undesirable practices of which arms manufacturers had been accused.

The result was that by the latter part of 1934 there was a considerable demand in Britain for a public enquiry into the arms industry and the case for its nationalization. At first the government did not seem at all inclined to yield to it. When Clement Attlee, leader of the Labour Opposition in the House of Commons, attacked the arms trade in November and proposed that Britain should set an example to the rest of the world by forthwith prohibiting private manufacture, he received short shrift from the government spokesman Sir John Simon. The opposition, said Simon, had compared the arms trade with prostitution and the slave trade, yet they wanted it to be nationalized. Did this mean that they thought the government ought to set up nationalized brothels, or that an earlier government ought to have nationalized the slave trade instead of suppressing it? As for the attacks made on arms manufacturers in recent books and pamphlets, these were not argument but propaganda, sometimes of a grossly misleading character. The objections to private manufacture listed in the report of the Temporary Mixed Commission of 1921 (and repeated in that of a committee appointed by the League of Nations Union in 1932) were not charges that had been investigated, but 'contentions set out on one side'.

However, the demand for an enquiry did not come only from men and women convinced by propagandist literature that arms manufacturers were immoral. Members of Parliament who thought an enquiry was overdue included a former Minister of Munitions, Dr Christopher Addison, and the leader of the Liberal Party, Sir Archibald Sinclair. Their arguments were founded not so much on morality as on expediency. Moreover, the experiences of 1914 to 1918 had convinced a great many people without strong political affiliations that war was a thoroughly bad way of settling disputes between nations. Many of them attributed the outbreak of hostilities in 1914 at least partly to the piling up of armaments by the great powers in the preceding years. They were unenthusiastic about the government's rearmament programme because they feared another arms race which might lead to another war. Whether nationalization would reduce or increase the risk of an arms race was a difficult question on which an enquiry might be expected to throw some light. Finally, there was something in the argument that arms manufacturers ought to be given a chance of replying to the charges brought against them.

In the light of these and other considerations, the government came to the conclusion that the case for an enquiry was, after all, too strong to be resisted. A fortnight after he had demolished Attlee's arguments, Simon confessed that he had failed to gauge the state of feeling in the House and had not said what he ought to have said. On 21 December the Prime Minister announced that a Royal Commission would be appointed to go into the whole question of the private manufacture and sale of arms.

The commission was appointed on 20 February 1935. It was headed by a retired Lord Justice of Appeal, Sir John Eldon Bankes; its six other members included the well-known journalist J. A. Spender, the novelist Sir Philip Gibbs, and Dame Rachel Crowdy, Chief of the Social Questions and Opium Traffic Section of the League of Nations. A distinguished academic lawyer, a prominent figure in the Co-operative movement, and a business-man who had served on a Royal Commission on the Coal Industry made up the number. Few, if any, of these people could be suspected of any marked prejudice in favour of the private manufacture of armaments, and the publicity given to the Craven-Spear correspondence as a result of the American enquiry made it almost certain that the proceedings would open in an atmosphere of considerable hostility towards the arms trade in general and Vickers in particular.

The commission began by drawing up long lists of questions to which the leading firms were asked to give written answers. Vickers devoted a great many man-hours to the preparation of detailed replies and of briefs for representatives who would afterwards appear as witnesses.[6] Imperial Chemical Industries replied in the same spirit to the questions sent to them. The documents drawn up by both companies provided a mass of information about their assets, turnover and profits, their holdings in subsidiary and associated companies and their relations with agents and competitors.

Imperial Chemical Industries was a company about which the public knew little when the enquiry began, except that (like E. I. du Pont de Nemours and Company) it was very big and drew part of its revenue from the manufacture and sale of explosives. It had been formed in 1926 on the initiative of Sir Harry McGowan—who had joined Nobel's Explosives Company in 1889 as an office boy—to unite Nobel Industries Limited with Brunner, Mond Limited, the British Dyestuffs Corporation and the United Alkali Company. The first of these four companies—originally called Explosives Trades Limited—McGowan had formed soon after the armistice by amalgamating forty firms which produced between them most of the high-explosive, gunpowder, detonators and small-arms ammunition manufactured in the United Kingdom. The second had been founded in 1873 by Ludwig Mond, an immigrant from Germany, and John Brunner, the British-born son of a Swiss father. Originally Brunner Mond had concentrated on the manufacture of soda, but in the course of the years it had absorbed four firms whose products included soap and a variety of chemicals. The third company McGowan regarded as an essential component of a chemical combine; the fourth was, until the amalgamation, Brunner Mond's chief rival. Interests acquired by Imperial Chemical Industries as a result of its absorption of Nobel Industries included joint control with De Beers Consolidated Mines of a South African explosives company which afterwards became African Explosives and Chemical Industries Limited.

On 1 May 1935, the commission held the first of twenty-two public sittings. These were supplemented by four sittings held in private for the purpose of hearing confidential evidence from Sir Maurice Hankey and Sir Eric Geddes and of receiving submissions from 'representatives of a British armament firm in regard to the activities in a foreign country of an employee of that firm'. The commissioners examined seventy-three witnesses, some of whom came forward at their own request while others were asked to do so. In addition to Hankey and Geddes, witnesses who offered testimony founded on special knowledge of problems of procurement and supply included Lloyd George and Christopher Addison. The chief producers of explosives, warships, guns, tanks and military or naval aircraft were represented by eight witnesses from Imperial Chemical Industries, six from Vickers and associated companies, seven from the Society of British Aircraft Constructors, two from William Beardmore and Company and one from Hadfield Limited. The War Office sent four witnesses, the Admiralty three, the Air Ministry four and the Foreign Office two. The commission also received submissions from the League of Nations Union, the National Peace Council, the Union of Democratic Control, the Independent Labour Party, the Communist Party of Great Britain and the British Movement against War and Fascism. Philip Noel-Baker, a well-known campaigner for the abolition of private manufacture, appeared on his own behalf.

Of the witnesses who favoured nationalization, Christopher Addison was perhaps the most persuasive. The substance of his case against arms manufacturers was not that they fomented war-scares or joined price-rings, but that they were inefficient and unreliable. Having served at the Ministry of Munitions as Parliamentary Secretary or Minister from its formation in 1915 until 1917, he could speak with authority of the performance of the national factories set up by the Ministry. But his interpretation of the figures on which he based his comparison between their achievements and those of the private sector was open to question, and his account of earlier events was not always accurate. His allegation that the private firms had let the country down in 1914 by promising more shells than they could deliver was flatly contradicted by the then Master-General of the Ordnance, who had the advantage over Addison of having attended the meeting at which the promises were alleged to have been made. The Master-General testified that the manufacturers, when asked whether they could increase their output, replied that they thought they could, but could make no promises. When told that his evidence differed from Addison's, he said simply: 'I was present at the meeting. I do not know what Dr Addison was doing then; at any rate, he was not there.'

The best-informed and most influential opponent of nationalization was Hankey. In his oral evidence and two written submissions he dealt with virtually the whole of the case for and against abolition from almost every conceivable aspect, moral and ethical as well as practical. The passage in the Covenant of the League of Nations which described the private manufacture of arms as open to grave objections had been inserted, he said, to please President Wilson; it did not represent the views of the British delegates to the Preliminary Peace Conference, or perhaps of all the American delegates. The Temporary Mixed Commission had sought to answer the question: 'What objections?' by listing those known to have been voiced by abolitionists. Hankey then set out, in minute detail, the evidence and the chain of reasoning which led him to conclude that the Temporary Mixed Commission had not investigated the charges but had merely recorded them.

In their report, issued in the autumn of 1936, the commissioners specifically endorsed that conclusion.[7] They added that very few occurrences which supported the allegations made against arms manufacturers had come to their attention. They had found no convincing evidence that British firms actively promoted war-scares or persuaded the authorities to adopt warlike policies. They did not believe that British firms tried to bribe representatives of their own government, but were aware of evidence that bribes had been offered to representatives of foreign governments. For that reason they did not approve of 'blanket' commissions which could be used to hide corrupt payments. They had come across only one piece of evidence which bore on the allegation that arms manufacturers had initiated false reports about the naval or military programmes of foreign countries—and even that, as the record shows, was of a very slight and inconclusive character. They had found no evidence that British firms had sought to influence public opinion through the control of newspapers at home or abroad, or that they had engaged in nefarious and clandestine activities 'connected with price-raising and other rings.' Legitimate price-maintenance agreements, they said, were sometimes disguised by 'a form of collusive tendering'. They did not feel able to pronounce upon the alleged existence of international armament trusts which raised the price of armaments sold to governments—but it is perhaps significant that they made no adverse comment on the arrangement between Vickers and the Electric Boat Company. They did, however, find fault with the 'frivolous and cynical' tone of some of Craven's letters. And Sir Philip Gibbs was not best pleased when Craven, on being asked whether he thought his wares more noxious than boxes of chocolates or sugar candy, replied: 'Or novels, no.'

On the general question of the morality of the arms trade, the commissioners said that the mere fact that profits were earned did not, in their opinion, make the private manufacture of arms objectionable. They saw no harm in the employment by arms manufacturers of retired officers or officials, but suggested that the permission of the appropriate minister should be obtained as a sop to public opinion.

A point not overlooked by the commissioners was that whether weapons of war served acceptable or unacceptable ends depended not on the morality of their manufacturers but on that of their recipients. It seemed to follow that control of the international traffic in arms was crucial from the point of view of statesmen who wished to preserve peace without committing themselves to the utopian ideal of a world totally disarmed. The commissioners noted that nearly forty nations had acceded to a convention of 1925 which sought to regulate the transfer of arms from one country to another, but that only about a third of them had ratified it and that nothing had been done to make it effective. The British government, they said, had complete control over the export of arms, but this did not apply to aircraft although it did apply to any armament they carried.

That, of course, was true inasmuch as an Order in Council of 1931 prohibited the export of arms from the United Kingdom except under licence. Similar steps were taken in the 1930s by other industrial countries. But such measures were not an effective substitute for undertakings to regulate the traffic in arms in accordance with an international convention. They could have provided the teeth and claws of such a convention. They did furnish governments willing to exercise their powers with a convenient means of doing so. But they did little to help ministers and officials to decide in what circumstances a licence should be granted or withheld. Their limitations were well shown by the difficulty experienced by the League of Nations in stopping a war between Bolivia and Paraguay. Although the League succeeded, after a long delay, in persuading not only its members but also the United States to ban the sale of war materials to both sides, the belligerents continued to receive supplies under existing contracts for some months after embargoes were imposed.

The commissioners also touched on the sale of surplus and used equipment. A spokesman for an American firm which held large stocks of weapons dating from the First World War had told the Nye Committee that a British company with which it was linked by an agency agreement claimed to have sold more than a hundred thousand rifles to China in 1931 and 1932. The British company was the Soley Armament Company, a respectable firm well-known to the British military authorities. In 1930 the War Office had made a contract with BSA Guns Limited which gave BSA the sole right to purchase and resell its surplus Lewis guns and 1914-pattern rifles. BSA had appointed Soley to act for them, and in 1931 Soley had enlarged their scope by acquiring the right to purchase and re-sell Hotchkiss guns and steel helmets. All these were legitimate transactions, calculated to reduce the burden borne by the taxpayer; but the commissioners viewed with evident disapproval the sale of surplus weapons to commercial firms. Their concern was understandable in view of the ease with which, according to evidence tendered to the Nye Committee, restrictions on the export of rifles and machine-guns could be evaded by dealers less scrupulous than BSA and Soley.

So much for the moral aspect. On the issue of expediency, the substance of the commission's findings was that the abolition of private manufacture would neither enhance the prospects of peace nor enable the nation to prepare more efficiently for war. In the past, the commissioners said, the Royal Navy had relied in peacetime on private enterprise for nearly all its ships and armaments except a few ships built in naval dockyards and some special requirements. The British Army had lived for some years after the First World War largely on stocks, but the time when it could afford to do that had passed or was passing. It relied on private firms for practically all its tanks, the greater part of its ammunition and certain other items such as gun forgings and mountings for heavy coast-defence guns. The Air Ministry was almost entirely dependent on private enterprise. For many years the policy of successive governments had been to keep alive an effective private industry to meet peacetime needs and provide a reserve of manufacturing capacity on which the nation could draw in time of war. If private firms were not used in peacetime and the armed forces became entirely dependent on government factories, the capacity to expand in time of war would be lost, or at best substantially reduced.

This argument leaned heavily, of course, on the assumption that in peacetime a nationalized arms industry would cater solely for the needs of the British government and would not accept or solicit orders from foreign governments. That assumption was still valid in 1936 and for many years thereafter. That a British government should expose itself to the hurly-burly of the market-place by entering the international arms trade in competition with Krupp and Schneider was unthinkable in the circumstances of the day. The commissioners drew the logical inference that any substantial departure from the existing system would do more harm than good. They recommended, however, that the government should make it clear to the general public that in time of war the private firms would be brought under strict control and would not be allowed to make excessive profits. They also recommended that an organization should be set up to regulate supplies and decide priorities.

Thus the battle between private enterprise and the champions of nationalization resulted in a victory for private enterprise. Even so, the victors did not escape without a few flesh-wounds. Vickers were closely questioned about their part in two transactions involving in the one case undoubted bribery, and in the other what seemed to the commissioners to have been intended bribery.

The first of these had been the subject of a scandal which erupted when two Japanese admirals were convicted in 1914 of accepting bribes in connection with orders placed with British firms. The affair had been widely publicized at the time, and questions had been asked about it in the House of Commons. No one working for Vickers in 1936 had any first-hand knowledge of the facts, and much of the information furnished to the commissioners came from contemporary newspaper reports. These and the company's records showed that in 1910 the Japanese government had placed an order with Vickers for the warship *Kongo* at a contract price of £2,307,100. The Japanese firm of Mitsui Bussan acted as sole agents for Vickers and were paid on a commission basis. Admiral Matsumoto, at the relevant time

Director of Naval Stores, was said at his trial to have been rewarded for placing the order with Vickers by receiving one-third of Mitsui Bussan's commission. The other Japanese admiral, whose name was Fujii, admitted in court that he received substantial sums from Vickers after reporting favourably on their estimate and specifications. He was said to have received also a more modest reward for ordering turbines for the *Hiyei.* Six payments made to Fujii between December 1910 and the summer of 1912 appeared from the records of the Barrow shipyard to have been debited to the cost of the *Kongo* and to have been authorized by Sir James McKechnie. Craven told the commissioners that he thought McKechnie ought to have been dismissed when the Japanese trial brought the matter to the attention of the board, and that doubtless he would have been dismissed if the outbreak of war had not made him indispensable. All this was very distressing, but Lawrence and his colleagues could scarcely be blamed for the failure of their predecessors more than twenty years earlier to send McKechnie packing in the midst of a national emergency.

The second transaction was more recent, and its relevance to the allegation that arms manufacturers were in the habit of bribing officials was not so clear. It came to light as the result of an invitation given by Craven to Dame Rachel Crowdy and Sir Philip Gibbs to visit Vickers House and spend as much time as they liked examining any papers they cared to see. After looking through hundreds of files they found a reference to an incident that had occurred in China in 1932. A demonstrator of tanks employed as an agent by Vickers had caused £50,000 to be added to a price. The money had not been paid and the man had been called home and dismissed. The commissioners, taking the view that the man had intended the £50,000 to cover a corrupt payment to some Chinese warlord, stated in their report that a British firm (which they did not name) deserved censure for conniving at a bribe. The directors of Vickers Limited believed that the inference drawn by the commissioners was unsound. In their view the man, whom they believed to have been suffering from delusions of grandeur, had intended the money for his own pocket. They recognized that they had made an error of judgement when they engaged him, but were sure they had not connived at corruption. They took counsel's opinion, and were advised that there was no evidence of intended bribery or of connivance by the board.[8]

But the general public, of course, did not know that. It did know that representatives of a British firm—which it guessed to be Vickers—had been interrogated at a private session about the activities of an employee in a foreign country, and that a firm which again it guessed to be Vickers had been reprimanded for conniving at a bribe. In any case the report was bound to leave some ends untied. The commissioners were not in a position to investigate charges brought against arms manufacturers who were not British subjects or domiciled in Britain, even though some of the charges had been widely publicized in books or pamphlets on sale in the United Kingdom. They could not, for example, throw any useful light on the extent to which the attitudes of French governments to problems of disarmament and rearmament had been affected by the opinions of members of the Comité des Forges known to have friends in high places and said to control at least two French newspapers. Nor could they go into the details of a scandal which had erupted in Rumania when Skoda's agent in that country was found to have secret military documents in his files and alleged to have distributed large sums in bribes. But perhaps more disturbing than any of this was the knowledge that restrictions on the international trade in arms had not prevented Bolivia and Paraguay from preparing for war, or China and Japan from buying large numbers of weapons and aircraft and large quantities of warlike stores from suppliers on both sides of the Atlantic.

Intended by its sponsor as a 'broom' to sweep the trenches, the Tommy gun acquired a doubtful reputation because of its use between the wars by gangsters and strike breakers

TEN

Rearmament

Immediately after the First World War, a number of people with a taste for the politico-strategic equivalent of science fiction predicted that the next great war would be fought in the Pacific Ocean and would start with an attempt by the Japanese to invade Australia and to seize British and American possessions in the Far East and the South-West Pacific.

Not everyone believed these predictions. The British naval authorities formally rejected a report which depicted Japan as a potential aggressor. The fact remains that the British government of the day, besides allowing the Japanese alliance to lapse under pressure exerted mainly by the United States but also to some extent by Australia and Canada, went so far as to assent in principle to the construction at Singapore of a base to which the main fleet could be sent in the event of trouble with Japan.

In fact, of course, the Japanese had no intention immediately after the war of invading Australia or of trying to seize Hong Kong, Malaya or the Philippines. Her statesmen still attached great importance to good relations with the West. She was pleased to have been awarded League of Nations mandates for the former German colonies in the Pacific north of the equator, but she valued still more highly her inheritance of the former German concession at Tsingtao, in the Chinese province of Shantung. Her annexation of Korea long before the war had made her not merely an insular but an Asiatic power. Her overriding aim—the purpose for which she had built up a powerful navy—was to establish a strong position in Continental Asia by developing her interests there and by exploiting the manifest inability of the ramshackle Chinese Republic to resist economic penetration. The question was how far she would be allowed to have her way by the well-entrenched powers whose Far Eastern interests included a big stake in China's foreign trade.

Japan became, and remained for more than a decade, a staunch upholder of the League of Nations. The loss of her alliance with Britain was a great disappointment to her, but she accepted with a good grace the provisions of the Naval Treaty of Washington. In the 1920s she followed the example of Britain and the United States by substantially reducing her military establishments.[1] In the context of the Washington agreements she deferred to American and Chinese wishes by relinquishing the Shantung concession, but she retained her economic interests in Manchuria and the right to maintain armed forces there, in the neighbourhood of the legation quarter at Peking and its communications with Tientsin, and in the International Settlement at Shanghai. In the same context she undertook to refer any dispute between her and the British Commonwealth, France or the United States to a conference of all four powers. Under the Nine-Power Treaty, also negotiated at Washington, all the powers with Far Eastern interests except Russia pledged themselves to respect the sovereignty, rights, interests and territorial integrity of China. Russia was not invited to Washington because the United States government did not enter into diplomatic relations with the Soviet Union until 1933.

In the light of these pledges, and in view of Japan's dependence on imports for all or most of her refinery products, bauxite, iron ore, raw cotton and natural rubber, the British gave only a low priority to the development of Singapore and the Western Powers saw no harm in supplying Japan with arms and military aircraft. When the British Foreign Secretary, Sir Austen Chamberlain, was asked by the Committee of Imperial Defence in 1925 for his views on the future of Singapore, he said that a major clash in the Far East seemed unlikely to come without warning, and that Britain and the United States ought in his opinion to be able to ensure, by presenting a united front, that Japan did nothing in China which was offensive to them. Doubtless Chamberlain was thinking of economic rather than military co-operation. Perhaps it is not surprising that he failed to foresee the difficulty both governments would have in bringing themselves to restrict trade with Japan in conditions very different from those of 1925.

The first unmistakable sign of an impending rift between

Japan and the Western Powers came when the Chief of the Japanese Naval Staff resigned as a protest against his country's adherence to the London Naval Treaty of 1930. The government's critics alleged that its civilian members had defied the constitution by disregarding his advice. They also accused the government of persistently subordinating the interests of the armed forces and of landowners and farmers to those of industrial magnates and the urban proletariat. Supporters of an influential and highly vocal patriotic movement called for a more equitable distribution of incomes and for pressure on the Western Powers to relax their grip on the lion's share of the world's raw materials.

Such was the state of affairs in Tokyo when, on 18 September 1931, a mysterious explosion wrecked a short stretch of the Japanese-controlled South Manchurian Railway. Shots were exchanged between Chinese troops and the Kwantung Army, as the force which the Japanese were allowed to maintain in Manchuria was called. The Chinese, alleging that the Japanese had staged the explosion in order to provoke a clash, appealed on 21 September to the League of Nations. Later they instituted a boycott of Japanese goods. This led to a riot and the landing of a Japanese expeditionary force at Shanghai, but the Japanese were soon persuaded by mediators to withdraw.

The Manchurian imbroglio had more serious consequences. The League of Nations was in no position to intervene in a conflict between China and Japan by banning the sale of arms to the disputants, especially as neither the United States nor the Soviet Union was a member of the League. On the contrary, the League found itself powerless to prevent China and Japan from substantially increasing their imports of war material in 1931 and 1932. Its attempts to settle the dispute were confined to a demand that the Japanese should withdraw their troops from Manchuria and the despatch of a fact-finding mission.

The Japanese could scarcely be expected to confess themselves in the wrong by ordering the Kwantung Army to withdraw at the behest of mediators who had yet to inform themselves of the bare facts of the dispute. In any case it seems unlikely in the light of subsequent events that any such order would have been obeyed. What did happen was that the Kwantung Army took it upon itself to seize strategic points throughout Manchuria and set up the puppet-state of Manchukuo. The League's fact-finding mission completed its labours in time to chronicle these events and record its opinion that Manchukuo was an artificial creation which owed its existence solely to Japanese intervention.

In the meantime popular opinion in Japan became so hostile to statesmen associated with the policy of friendship with the West that the government lost the power to control events at home or abroad. Armed malcontents roamed the streets of Tokyo unchecked, and at times actively supported, by members of the fighting services. Two prominent industrialists and a Prime Minister were assassinated. In the spring of 1933 the Emperor, acting as always on advice, turned his back on the party political system and appointed a non-party government acceptable to the armed forces. The new administration formally recognized Manchukuo, took Japan out of the League of Nations, and initiated a programme of rearmament based on the principle that the fighting services should, as far as possible, be equipped with weapons of Japanese design and manufacture.

The British Chiefs of Staff described events in China in 1931 and the early part of 1932 as 'the Writing on the Wall'. They urged the government to abandon the assumption that there would be no great war involving the British Empire for at least ten years, and to start providing for purely defensive commitments without delay. The government agreed to cancel the ten-year rule and sanction completion of the naval base at Singapore and its fixed defences by 1936, but was unwilling to do more while the outcome of the Disarmament Conference remained uncertain.

The next milestone on Britain's long road to rearmament came in February 1934, when the Defence Requirements Committee appointed by the government in the previous November submitted its report. The committee based its findings on the assumption that, although the immediate danger lay in the Far East, the ultimate potential enemy was Germany. There was no reason to suppose that Hitler contemplated an attack on any part of the British Empire, and we now know that war with Britain during the next few years formed no part of his plans. Nevertheless the British might be called upon to honour their Locarno pledges if his expansionist policies brought him into conflict with the French. By building a strong air force the Germans might, it was thought, put themselves in a position to go to war in 1938 or 1939. The committee recommended that the metropolitan air force should, as a matter of first importance, be brought up to its planned strength of 598 first-line aircraft. It also recommended a modest expenditure on the coast defences and on schemes of local naval defence. But its most important and most controversial proposal was that the War Office should prepare for despatch to the Continent within a month of the outbreak of war an expeditionary force of four infantry divisions and one cavalry division, augmented by a tank brigade, two air defence brigades, and an air component to be drawn from the metropolitan air force. The role of the expeditionary force would be to co-operate with Continental allies in securing the Low Countries, so that bases for an advanced air striking force and early-warning posts for the air defence of the United Kingdom could be established there. The committee's proposals would involve a capital expenditure of £71,000,000, to be spread over the next five years.

The government agreed that the metropolitan air force must be strengthened, but turned down the rest of the committee's programme on the ground that its cost was beyond the nation's means. The Chancellor of the Exchequer, Neville Chamberlain, argued that, if the object was to convince Hitler that Britain would not stand by while he helped himself to *Lebensraum* at the expense of Eastern European countries allied to France, the government could attain it more cheaply and more certainly by outmatching him in the air than by equipping an expeditionary force which would in any case be small by Continental standards. Accordingly the government lopped about a third from the Defence Requirements Committee's estimate, reduced the army's allocation by fifty per cent, and sanctioned a scheme of air expansion designed to give the metropolitan air force a first-line strength of 960 aircraft by the end of 1938 or early in 1939.

By the time these decisions were made, well over two years had elapsed since the ten-year rule was cancelled. In the meantime—and indeed for some time afterwards—arms manufacturers received only modest orders from the British government. At the same time the demand for British-built merchant shipping continued to fall sharply. Some firms, which could have made valuable contributions to the rearmament drive if it had started as soon as the Chiefs of Staff uttered their warning, were unable to make ends meet, with the result that their productive capacity was lost to the nation, at least for the time being. A notable example was the old-established firm of Palmers, which had built the first warship armoured with rolled-steel plates and whose yard at Jarrow had been building ships for the Royal Navy since Nelson's day. In 1932 the firm, already shorn of its steelworks and no longer owned by the Palmer family, was obliged to part with its shipbuilding interests to a company which kept the shipyards open only for repairs.

The naval strengths of the leading maritime powers were governed in 1934 by the Washington Naval Treaty and the London Naval Treaty of 1930, but all restrictions on new construction would be swept away at the end of 1936 unless a new treaty could be negotiated in the meantime. The gist of the arrangements made in 1930 was that the signatories pledged themselves to build no capital-ship replacements before 1937 and that Britain assented to an upper limit of fifty cruisers and accepted parity in submarines and destroyers with Japan and the United States. According to a calculation made before Hitler came to power, to be able to meet the Japanese on equal terms at their chosen moment Britain would need to station in the Far East about twelve capital ships, five aircraft carriers and forty-six cruisers. The Admiralty hoped, at the cost of reducing the Home Fleet to little more than a token force of three capital ships and four cruisers, to be able to provide a fleet of approximately that size by the time it was likely to be needed, but whether it would be qualitatively a match for the Japanese fleet was another matter. With the exception of the fourteen-year-old battlecruiser *Hood* and the seven-year-old battleships *Nelson* and *Rodney*, all the capital ships at the navy's disposal in 1934 were at least seventeen years old. The fifty cruisers envisaged in the London Naval Treaty were expected to be ready by 1936, but not the whole of the destroyers and submarines needed to give numerical parity with Japan and to complete a balanced fleet.

Even on the assumption that the German Navy would present no serious threat to the British Empire or to ocean trade in the foreseeable future, the outlook in 1934 was therefore far from reassuring. The Royal Dockyards could do something to meet demands for cruisers, destroyers and submarines, but they were not capable of building the five up-to-date capital ships whose construction had been planned before the London Naval Treaty was signed and would start when it expired. At least six firms—John Brown, Cammell Laird, Fairfield, Harland and Wolff, Swan Hunter and Vickers-Armstrongs—were qualified to build the largest ships, but the supply of guns, gun-mountings and fire-control equipment would almost certainly prove the limiting factor. Vickers, William Beardmore and Company and the Royal Arsenal at Woolwich had managed even when orders from the Admiralty were at their lowest ebb to retain some capacity for the manufacture of naval ordnance, but it was fairly small even in relation to needs felt before the rearmament drive began. For gun-mountings and fire-control equipment the Admiralty would rely entirely on the private sector. Vickers were the chief (and in 1934 the only) suppliers of gun-mountings—a position they owed to their determination to preserve at least some of their plant when competitors were scrapping theirs, and to the large liquid reserves and ability to attract orders from foreign governments which had enabled them to meet the cost of doing so. They were also the most important of the four firms which ranked in 1934 as suppliers of fire-control equipment.

Only a very moderate use was made of these resources during the next two years. Until war was imminent the government clung to the principle that rearmament must not be allowed to interfere with normal trade. The reluctance of ministers to ban exports of arms and restrict commercial transactions so soon after the country had emerged from a disastrous slump was understandable, but it was one of a number of aspects of their policy which helped to give potential enemies the impression that Britain was not taking rearmament very seriously. In 1934 the Barrow shipyard was half empty although three large passenger liners in addition to two destroyers and four submarines were under construction there. The gun-mounting department at Elswick was working at half capacity, and the corresponding department at Barrow was busy with mountings for Denmark, Finland, Portugal and Siam but had some capacity to spare. A year later the Barrow shipyard was still only about half full, but the Vickers-Armstrong naval yard on the Tyne had reopened in the meantime.

However, this did not mean that the Admiralty's suppliers had no anxieties about their ability to cope with demands that would be made on them when the new capital-ship programme and the modernization of existing capital ships were put in hand. Each of the five 35,000-ton battleships of the *King George V* class would carry ten 14-inch guns in quadruple and twin mountings. Quadruple mountings for 14-inch guns were extremely massive yet highly complex structures which took about three years to make, occupied a great deal of space and required for their installation cranes capable of lifting well over fifteen hundred tons. Towards the end of 1935 Vickers undertook in consultation with the Admiralty to spend a quarter of a million pounds on a new gun-mounting shop at Barrow and another quarter of a million on machine-tools. A further quarter of a million would have to be spent at Elswick, and the English Steel Corporation needed nearly a million, chiefly to expand its capacity to meet naval requirements. In the context of these arrangements, the Admiralty promised orders for capital-ship mountings for seven ships in the next four years.

There remained the problem of recruiting labour to match. This was always difficult at Barrow, because the place was remote from any large centre of population and there was no alternative employment to attract people to the neighbourhood. But even where skilled tradesmen were more plentiful, as on Tyneside and Clydeside, difficulties were apt to arise because some jobs offered better prospects of steady employment than others and were therefore more attractive. By the end of 1937 Vickers-Armstrongs were falling behind schedule with most of their armament work, largely though not solely because of manpower problems.[2]

Britain's attempt to outmatch Germany in the air landed the government in trouble almost from the start. When the first air expansion scheme—called Scheme A—was introduced in the summer of 1934, the Germans were expected to complete Stage 1 of *their* air expansion programme at the beginning of October 1935. This would give them a first-line strength of 576 aircraft. Information received during the next few months indicated that Stage 1 was going according to plan. If Stage 2 went equally well, the Luftwaffe would attain by the autumn of 1936 a first-line strength of 1,368 aircraft if immediate reserves were counted as part of the first line, or 1,026 if they were excluded. Winston Churchill, and also the Foreign Office, thought the Luftwaffe might be able to improve considerably on these figures. Hitler went so far as to claim in the spring of 1935 that it was already as strong as the Royal Air Force. In any case, clearly Scheme A would not provide the parity with Germany which was the government's declared aim.

It was soon succeeded, therefore, by a new scheme, Scheme C. This was intended to provide 1,512 first-line aircraft by the spring of 1937. Framed at a time when a fresh attempt to persuade France and Germany to enter into an arms pact was in prospect, it showed all too clearly that the government was more concerned with the impression its announcement of the scheme might be expected to make than with realistic preparations for war. Well over a third of the aircraft envisaged in the scheme were to be light or medium bombers. The light bombers would not be able to reach worthwhile targets in Germany from United Kingdom bases, and no satisfactory medium bomber had yet been chosen when the scheme was adopted. Like its forerunner, Scheme C made no provision for stored reserves.

By the early part of 1936 even the most sanguine members of the government had to admit that the chances of shepherding the French and the Germans into an arms convention were negligible. Meanwhile relations with Italy had deteriorated as a result of Mussolini's invasion of Abyssinia. In February the government abandoned the unconvincing Scheme C in favour of the first air expansion scheme which made genuine provision for needs likely to arise in war. Scheme F was designed to give by the spring of 1939 a metropolitan air force of 1,736 first-line aircraft, backed by immediate and stored reserves equivalent to 225 per cent of first-line strength. By the same date the first-line strength of units abroad would rise to 468 aircraft, as compared with 292 under Scheme A; that of the Fleet Air Arm to 312 as compared with 213. So completion of the scheme would give the air forces at home and overseas about 8,000 aircraft of first-line type in squadrons, workshops and stored reserves.

The production of enough aircraft to enable that figure to be attained in little more than three years was not expected to be easy. Some twenty aircraft manufacturing firms were regarded by the Air Ministry as worthy of receiving the specifications by which it made its requirements known.* Most new designs were

* These were: Airspeed (1934) Limited; Sir W. G. Armstrong-Whitworth Aircraft Limited; the Blackburn Aircraft Company Limited; Boulton Paul Aircraft Limited; the Bristol Aeroplane Company Limited;

initiated by one or other of some fourteen of these firms. Two of the firms—the Bristol Aeroplane Company and de Havilland—made both airframes and aero-engines. Aero-engines were also manufactured by Armstrong Siddeley Motors Limited, D. Napier and Son Limited and Rolls-Royce Limited. Rolls-Royce and Bristol were responsible for most new designs, but a substantial manufacturing potential existed at Armstrong Siddeley and Napier and outside the industry.

The authorities had already discussed with a number of firms a plan for the wartime production of aircraft and aero-engines in 'shadow factories' to be established by some of the leading manufacturers of motor-cars. On the introduction of Scheme F they decided to start putting it into effect without waiting for the outbreak of war or the declaration of a national emergency. For all practical purposes the principle that rearmament should not be allowed to interfere with normal trade was thereupon abandoned so far as the aircraft industry was concerned, but not otherwise.

Scheme F was adopted at a moment when new aircraft of revolutionary design were under development but not expected to come forward in large numbers until the scheme was approaching its final stages. To bridge the gap, to encourage established firms to expand and to enable the sponsors of the shadow factories to gain experience, the authorities decided to place substantial orders for Bristol Mercury VIII air-cooled engines, Fairey Battle light bombers and Bristol Blenheim medium bombers. In 1936 about a thousand Battles were ordered from Fairey Aviation and the Austin shadow factory, some fifteen hundred Blenheims from the Bristol Aeroplane Company, A. V. Roe and Company and the Rootes shadow factory. The Blenheim,

---

the Fairey Aviation Company Limited; General Aircraft Limited; the Gloster Aircraft Company Limited; Handley Page Limited; the de Havilland Aircraft Company Limited; the Hawker Aircraft Company Limited; the Heston Aircraft Company Limited; Miles Aircraft Limited; Percival Aircraft Limited; A. V. Roe and Company Limited; Saunders-Roe Limited; Short Brothers (Rochester and Bedford) Limited; the Supermarine Aviation Works (Vickers) Limited; Vickers (Aviation) Limited; and Westland Aircraft Limited. Airspeed, General Aircraft, Heston Aircraft, Miles Aircraft and Percival Aircraft were chiefly concerned with aircraft not of first-line type. Saunders-Roe specialized in flying boats and civil transport aircraft.

developed as a private venture, was extremely fast for an aircraft of its type when it first appeared in 1935; by the standards of 1939 and 1940 its performance was only moderately good. The Battle was obsolescent by the spring of 1939, but had to be used by the Advanced Air Striking Force in 1940 because no acceptable replacement was available.

These were essentially stop-gap orders. Scheme F called for a large number of medium bombers. The authorities had to find these aircraft, even at the cost of deeming the Battle to be a medium bomber. They recognized that ideally the best solution might be to forget about medium bombers and concentrate on building up a strong force of the new Handley Page Hampdens and Vickers Wellingtons, but they were unwilling to risk ordering these untried aircraft in large numbers and entrusting their manufacture partly to shadow factories. They also recognized that a stop-gap medium-to-heavy bomber would be needed if the Hampden and the Wellington turned out badly or if their production was delayed. Armstrong-Whitworth had prepared drawings in 1934 of a medium-to-heavy bomber for the Czechoslovakian Air Force. Two prototypes of a version called the Whitley had been built for the Royal Air Force, and 80 of these aircraft had been ordered as an exceptional measure before the prototypes were tested. On the introduction of Scheme F the authorities increased their order for the Whitley by 240 aircraft. They hoped in 1936 to replace the Whitley within five years by a heavier aircraft built to the specification that gave rise to the Avro Manchester, the Handley Page Halifax and the Short Stirling; in the outcome it remained in production until 1943 and was replaced by the Lancaster. Meanwhile the Royal Air Force went to war with a decidedly mixed bomber force of Battles, Blenheims, Hampdens, Wellingtons and Whitleys. The most advanced of these aircraft was the Wellington, designed by Barnes Wallis of Vickers Aviation in the light of his experience with the airship R.100.

So far as the fighter force was concerned, the British took their first step towards a new era of high-performance monoplanes in 1930, when the department of the Air Ministry concerned with supply and research suggested that manufacturers might be asked to apply to the design of landplanes lessons learnt from the record-breaking seaplanes of monoplane configuration designed by R. J. Mitchell of Supermarine Aviation for the Schneider Trophy contests. A specification was drawn up, but it called for a low landing speed and a maximum speed of only 250 miles an hour, as compared with the 329 miles an hour attained by the Supermarine S.6 in 1929 and the 407 miles an hour soon to be attained by the S.6b. Eleven entries submitted by seven firms included designs for monoplane fighters by Mitchell and by Sydney Camm of Hawker Aircraft. Camm's design was not accepted. A prototype was built to Mitchell's, but it proved so unsatisfactory that Supermarine rejected proposals from the Air Ministry for its modification and development. Both Camm and Mitchell then designed aircraft which embodied their own ideas of what a high-performance monoplane fighter ought to be, and the Air Ministry drew up new specifications to fit them. Camm's Hawker Hurricane first flew on 6 November 1935. Mitchell's Supermarine Spitfire made its maiden flight on 5 March 1936, not much more than a year before its designer died of lung cancer. The Hurricane was some 30 miles an hour slower, the Spitfire fractionally faster, than Willy Messerschmitt's Me. or Bf. 109, which also made its début in 1935.

Both the Hurricane and the Spitfire were private ventures in the sense that neither was designed to meet a specification existing when the drawings were prepared. Both designers, however, kept in close touch with the Air Ministry at the technical level and received useful guidance about operational requirements. Squadron Leader R. S. Sorley, of the Operational Requirements Branch of the Air Staff, played the chief part in persuading all concerned that the new fighters should carry eight ·303-inch Browning machine-guns apiece. The case for 20-millimetre Hispano cannon, mounted in the fuselage or possibly below the wings, had been considered but turned down in 1934. Wing-mounted cannon became an attractive possibility when the wings of the Hurricane and the Spitfire were seen in 1935 to be extremely strong, but the evolution of suitable mountings took so long that cannon Hurricanes were not used operationally before 1941. A cannon Spitfire was tried out early in 1939, and thirty of these aircraft were delivered in August 1940. A turret Fighter, the Defiant, was developed between 1935 and 1939 as an insurance against failure of the Hurricane and the Spitfire to come up to expectations.

Popular legend has depicted Mitchell as devoting his last few years of life to the design and development of the Spitfire because his awareness of what was going on in Germany convinced him that only prompt re-equipment of Britain's fighter force with fast monoplanes could save the country from the wrath to come. This interpretation of his motives has as firm a basis of truth as any imaginative writer or film director could hope to find, although Mitchell's contacts with the German aircraft industry were not, perhaps, as close as those cultivated by Supermarine's test pilot, Joseph Summers. (It was, for example, not Mitchell but Summers who reported after one of many visits to Germany that a particular factory seemed to be turning out 300-mile-an-hour bombers at the rate of two a day.) A tendency to equate German rearmament with hostile intentions towards Britain was not confined, however, to Mitchell and Summers or to visitors to German aircraft factories. Sir Robert McLean, until the autumn of 1938 Chairman of Supermarine and also of Vickers Aviation, was convinced that war with Germany was inevitable. '"*Der Tag*", as they used to say in 1914,' a director of Vickers wrote to Craven in March 1936, 'is now fixed for three years hence.'

Nor was the belief that the Luftwaffe was being expanded for the purpose of making war on Britain confined to aircraft and weapons manufacturers and designers. Anthony Eden, at the relevant time Lord Privy Seal and soon to become Foreign Secretary, felt his heart fill with 'grim foreboding' when Hitler claimed in 1935 that the Luftwaffe was already as strong as the Royal Air Force.[3] Air Chief Marshal Dowding, the first Commander-in-Chief of Britain's air defences under the system introduced in 1936, never doubted that sooner or later he would have to beat off German bombers. Squadron Leader Sorley shared McLean's opinion about the inevitability of war with Germany. So, it can safely be asserted, did most other members of the Air Staff. Moreover, they assumed not only that sooner or later Germany would make war on Britain but also that hostilities were more likely than not to start with air attacks on the United Kingdom. The Hurricane and the Spitfire were developed for the specific purpose of meeting such attacks.

This was a wise precaution. The fact remains that no evidence has come to light which suggests that at any time between 1935 and Chamberlain's announcement of solidarity with Poland in

1939 the Germans either contemplated the launching of such an offensive, or intended to go to war with Britain unless they were forced to do so. The knock-out blow for which the Luftwaffe was supposed to be preparing before World War II belongs to the same category of hypothetical dangers as the bolt from the blue for which the German Navy was supposed to be preparing before World War I. All the plans of conquest disclosed by Hitler to his service chiefs in the 1930s were consistent with the politico-strategic doctrine outlined in *Mein Kampf*. This was that Germany must at all costs avoid having to fight on two fronts. Above all, she must take care not to make an enemy of Britain, as she had done before 1914 by challenging her naval supremacy. *Lebensraum* must be found in Eastern Europe for the privileged inhabitants of a future Greater Germany, but not at the cost of a two-front war. Invariably the service chiefs responded to these expositions by drawing attention to the danger of war with the European democracies; invariably Hitler's rejoinder was that the European democracies would not fight. If they did oppose his wishes, he said, he would attack them suddenly and ruthlessly, but he did not expect to have to do so before 1943.

Hitler was not, to say the least, a stickler for the truth. But actions speak louder than words. The Germans went to great pains between Hitler's advent to power and the outbreak of war to build up their armed forces. They did not provide themselves with the large fleet of ocean-going submarines which would have been their strongest weapon against the British, or with transports and landing craft suitable for an expedition to the British Isles. Even the much-feared Luftwaffe was not really suitable for an offensive against the United Kingdom from German bases. Its true role was direct and indirect support of land forces. This included the bombing of objectives behind the lines, but not strategic bombing in the special sense in which the British used the term. As both sides were to find out later, the kind of bomber offensive the British had in mind before the war was virtually impossible without resources which neither they nor the Germans possessed in 1939.

On the whole, the British were very well informed about the potential enemy's first-line strength, but they tended to overestimate the output of the German aircraft industry and therefore to impute to the Luftwaffe much larger reserves than it possessed.

Production of all types, trainers included, rose from a monthly average of 31 in the year of Hitler's accession to power to 426 in 1936. For the next two years it remained almost stationary because additions to floor space and growing experience were offset by the retooling of factories for large-scale production of new aircraft tested in 1936. These included the Junkers Ju. 88, Dornier Do. 17 and Heinkel He. 111 bombers which formed the backbone of the German bomber force in 1940. The Junkers Ju. 87 dive-bomber was also tested in 1936. There was some doubt about the role for which it was best fitted, and output was modest.

The British faced very similar problems. Production of the Hurricane went fairly smoothly because the aircraft was comparatively easy to produce and because Hawker Aircraft were outstandingly successful manufacturers of fighter aircraft and could afford to base their plans on the assumption that they would receive a large order. Supermarine Aviation, although a wholly-owned subsidiary of Vickers Limited since 1928, was a small firm with no experience of large-scale production. Their initial order under Scheme F was for 310 Spitfires. This was followed by an order for 168 Walrus amphibians for the Fleet Air Arm. Arrangements were made for three-quarters of the components of the Spitfire to be built by sub-contractors. The technical staff of Supermarine had therefore to divide their time between visits to sub-contractors confronted with unfamiliar tasks, and the supervision of production and assembly in the firm's own factories at Eastleigh and Woolston. Six Spitfires could have been completed by the end of 1937 if wings had been available, but they were not.

In the meantime new preoccupations were added to the anxiety felt by the authorities about Spitfire production. By the time Scheme F was under way, the advent of radar (then called RDF) made an effective system of air defence a practical possibility—but only if the fighters, guns and searchlights needed to provide the teeth and claws of such a system could be provided. All the air expansion schemes, Scheme F included, put more emphasis on bombers than on fighters, and such anti-aircraft artillery and searchlight formations as existed in 1936 were grossly under-equipped and under-manned. In the summer of that year a review of air defence problems in the light of changed conditions, and a tour of inspection of vulnerable objectives by army and air force officers, led to the conclusion that ten times as many guns and twenty times as many searchlights as were then available in the whole of the United Kingdom were needed. These would have to be supplemented by weapons suitable for employment against low-flying aircraft. Soon afterwards the Air Ministry reported that the Luftwaffe had completed Stage 2 of its programme and could be expected to attain a first-line strength of 1,500 aircraft by the spring of 1937. Information from reliable sources indicated that thereafter it would continue to expand until a first-line strength of approximately 4,000 aircraft was reached.

This was an extremely accurate forecast of the first-line strength of the Luftwaffe on the outbreak of war. Ministers found it doubly disquieting because they did not know that the Germans were building up an impressive first line at the cost of allocating very few aircraft to stored reserves. The outlook seemed even gloomier when a committee of experts, invited to draw up an 'ideal' scheme of air defence regardless of considerations of supply, reported early in 1937 that many more fighters were needed than Scheme F would provide.

The authorities knew, of course, that considerations of supply could not really be left out of account. The limiting factor must always be the number of fighters the aircraft industry could produce. In 1937 it was employing little more than a quarter of the workforce employed at the height of World War I, although the number of man-hours needed to build an aircraft had increased tenfold. What the authorities might have to disregard, and ultimately did disregard, was the financial aspect. A calculation made in the spring of 1938 suggested that, by working double shifts and refusing all outside orders, the industry should be able to turn out 4,000 aircraft in the next twelve months and 8,000 in the following twelve.

With these figures before it, the government agreed on 17 April 1938 to replace Scheme F by Scheme L. The new scheme set a target for the metropolitan air force of 2,373 aircraft in first-line squadrons by the spring of 1940. Reserves estimated to cover sixteen weeks' wastage in time of war for fighter and general-reconnaissance squadrons and nine weeks' wastage for bomber squadrons would be provided. Bombers would still outnumber fighters by two to one, but the fighter force would be nearly 50 per cent stronger than under Scheme F.

Scheme F was thus superseded when it still had a year to run. About 4,500 of the 8,000 aircraft envisaged in the scheme had been delivered, but only some 1,500 of these were of up-to-date design. Supermarine had built thirty-five Spitfires but received only twelve complete sets of wings from their sub-contractors. Only one of the 310 Spitfires ordered under Scheme F had been completed when Scheme L brought an order for another 200, and even that could not be delivered until later in the year because last-minute modifications were ordered. The purchase of American aircraft was considered, but American manufacturers were found to have few military aircraft to offer which compared favourably with their British and German counterparts. Orders were, however, placed in June for 250 Lockheed Hudson bomber-reconnaissance aircraft and 200 North American Harvard trainers. Early in 1939 another 200 Harvards were ordered.

However, from the beginning of May the output not merely of Supermarine but of the aircraft industry in general showed a decided upward trend. One reason was the lifting of financial restrictions on the placing of orders. Another was that the benefit of the shadow factories and of government-financed additions to the productive capacity of established firms was beginning to be felt. By October Hurricanes and Spitfires were coming forward at the rate of twenty-five and thirteen a month respectively. By the end of the year Supermarine were marginally ahead of their production schedule and had started to make wings for Spitfires in their own factories. Altogether about 2,800 aircraft were delivered in 1938.[4] About half were trainers or other aircraft not of first-line type. Nearly 700 were light or medium bombers and some 370 were fighters.

In the meantime the Munich crisis brought an emergency deployment of Britain's air defences. Twenty-nine fighter squadrons were deemed mobilizable. Five were equipped with Hurricanes, the rest with biplanes of which the 250-mile-an-hour Gloster Gladiator was the best. There were no mobilizable Spitfire squadrons and no stored reserves of fighter aircraft. Germany was nothing like ready for a European war, but a bomber force of more than a thousand serviceable aircraft gave the Luftwaffe a formidable appearance.

Soon afterwards the Minister for the Co-Ordination of Defence, Sir Thomas Inskip, recommended that fighters rather than bombers should have first claim on productive capacity. Scheme M, adopted in November as a follow-up to Scheme L and intended for completion in 1942, reduced the planned proportion of bombers to fighters from 2:1 to 1.7:1. In the outcome, the proportion fell in 1939 to 1.4:1. Altogether the British produced about 8,000 aircraft in that year, the Germans about a thousand more. In each case about half were of first-line type. Orders for Canadian-built Hampdens, Stirlings and Hurricanes were placed in November 1938, but deliveries did not begin until after the outbreak of war.

One of the most spectacular features of the rearmament era was the failure of the French to match these efforts. At the beginning of the 1930s France possessed the strongest military air force in the world. In 1933 the Armée de l'Air was placed under an Air Ministry independent of the War Office, and a system was introduced by which operational squadrons would be grouped in wartime in formations allocated to Army Groups or their equivalent. The Naval Air Service remained outside the system, and the Ministry of Marine was responsible for choosing its equipment.

To offset the first stage of the Luftwaffe's air expansion programme, the French Air Ministry ordered in 1935 a large number of Dewoitine 510 fighters. The Dewoitine 510 was a good aircraft in its day, but it was superseded in the course of the next few years by the 326-mile-an-hour Dewoitine 520. The Dewoitine 520 had yet to make its maiden flight when, in 1936, the independence of the many small firms which made up the aircraft industry was threatened by the advent to power of a Popular Front government committed by its election pledges to far-reaching measures of nationalization. This might not have been disastrous if the new government had not fallen foul of the extreme Left by refusing to commit the country to active support of the Leftist cause in the Spanish Civil War. The outcome was that the industry was disrupted by political and industrial strife at a time when output ought to have been increasing by leaps and bounds.

In 1938, under a different government, a fresh attempt was made to reorganize the industry and put the Armée de l'Air on a sound footing. The immediate aim was a first-line strength of 2,500 aircraft. An output of 200 aircraft a month was planned, and orders were placed in the United States for Curtiss Hawk fighters for the Armée de l'Air and a few Chance-Voght dive-bombers for the navy. Later, more Curtiss Hawks and some Douglas DB-7 and Martin Maryland medium bombers were ordered. Peacetime output did not come up to expectations, but was better for fighters than for bombers. On the outbreak of war the Armée de l'Air had in metropolitan France about 325 Morane-Saulnier 406 and Curtiss Hawk single-seater fighters; some 70 Potez 631 heavy fighters; about 400 LeO 451 and obsolescent Amiot 143, Bloch 200 and Bloch 210 bombers; and a miscellaneous collection of reconnaissance and observation aircraft.[5] The Naval Air Service had some 350 aircraft, but only a few were of recent design.[6]

In everything but aircraft the French Navy was well equipped. Besides stationing three battleships, a seaplane carrier, ten cruisers, forty-eight destroyers and fifty-three submarines in the Western Mediterranean and small cruiser or destroyer forces at the Channel ports, in Western France, on the west coast of Africa and in the Far East, the French were strong enough on the outbreak of war to base on Brest a striking force for use against commerce raiders in the Atlantic. This consisted of the modern battlecruisers *Dunkerque* and *Strasbourg*, the aircraft carrier *Béarn*, three cruisers and ten destroyers. The 35,000-ton battleships *Jean Bart* and *Richelieu* were approaching completion.

The German Navy was in a far less favourable position. Its long-term programme was based on the assumption that there would be no confrontation with Britain before 1944 or 1945. A short-term programme, intended as an insurance against the risk of earlier hostilities, envisaged war with France alone. On the outbreak of war the navy possessed only twenty-six ocean-going and thirty smaller U-boats, of which forty-six were fully operational. Apart from torpedo boats, motor torpedo boats, mine-layers, minesweepers and two old battleships unsuitable for use outside the Baltic, its only surface warships were the battle-cruisers or fast battleships *Scharnhorst* and *Gneisenau*, six cruisers, seventeen destroyers, and the pocket battleships *Admiral Scheer*, *Deutschland* and *Graf Spee*. The last were essentially commerce raiders. The Royal Navy and the Commonwealth navies had in commission twelve capital ships, six aircraft carriers, two seaplane carriers, fifty-eight cruisers and more than two hundred destroyers, sloops or corvettes. These figures did not include any battleships of the *King George V* class or carriers of the *Illustrious* class, none of which would be ready before 1940.

The French Army was well provided with weapons capable of a formidable weight of fire. A four-year plan designed to bring its equipment up to date was put in hand soon after Hitler's remilitarization of the Rhineland. During the next two and a half years the legislature voted sums equivalent to some £227,000,000. Large numbers of field, medium and heavy guns left over from the First World War were brought out of store and modernized. New projectiles designed by Edgar Brandt, inventor of the Brandt mortar, increased the range of the field and medium guns. At the end of the four years the French had at their disposal well over seven thousand 75-millimetre and 105-millimetre guns and not far short of four thousand heavier pieces suitable for the reduction of permanent defence works.[7] Block-busting projectiles for the heavy artillery had yet to be delivered on the outbreak of war, but even without them the guns were capable of doing a great deal of damage. The French were, however, short of up-to-date anti-aircraft guns, and much of their artillery would be handicapped in a war of movement by a lack of mobility which reflected the preoccupation of the High Command with linear defence.

Contrary to popular belief, the French Army was also well provided with modern tanks. In 1939 and 1940 about 600 old Renault tanks were deployed for the local defence of airfields. This may have been one cause of a widely-held belief that the French had few or no tanks comparable with the German PzKw I, II, III and IV. Another was the tendency of French generals after the fall of France to attribute the army's shortcomings to inadequate equipment. In fact, the Renault, Hotchkiss and Somua factories and the Forges et Chantiers de la Mediterranée produced under the four-year plan more than 3,400 tanks of excellent quality, all built to designs approved in 1935 or 1936.[8] Most were armed with weapons capable of knocking out any German tank. The FCM Char B was acknowledged to be the best heavy tank in production in any country. Because the High Command gave priority to infantry support, speed and range were, however, subordinated to firepower and the capacity to withstand punishment in the design of all the French tanks. Class for class, they were therefore slower and needed more frequent refuelling than their German counterparts, although they were more thickly armoured and in most cases better armed.

*

The British Army was in many ways the antithesis of the French. France maintained a large standing army recruited by conscription and intended primarily for a static role. The British Army was small, professional and intended to be highly mobile. Although France was, like Britain, a colonial power, the primary task of her armed forces was to defend the homeland. The British relied for home defence chiefly on the navy. The French Navy's most important commitment was to secure command of the Western Mediterranean so that troops could be moved from North Africa to metropolitan France in the event of a European war.

After the First World War there was a good deal of controversy in British military circles about the army's future role. One point on which almost everyone could agree was that it should be well provided with mechanically-propelled vehicles as a step towards the mobility needed to make up for lack of numbers. The cavalry was understandably reluctant to lose its horses for tanks or armoured cars. But there could be no valid objection to the mechanization of Royal Army Service Corps, artillery, engineer and signal units and first-line transport—except that it cost money.

Here no serious problems of design or production were involved. Furnishing the army with lorries was a task well within the competence of the motor industry. Finding a tractor capable of pulling a field gun across muddy or broken ground was not quite so easy, but a satisfactory tracked vehicle called a dragon was evolved. The limiting factor was finance. Until 1935 the cost of all new equipment had to be defrayed from exiguous annual budgets. This restricted the number of wheeled vehicles ordered between 1923 and 1932 to an average of about 500 a year.[9] About 120 dragons were ordered from Vickers-Armstrongs in the early 1930s, but these did not go far to equip the artillery of five (afterwards six) field divisions of the Regular Army and twelve (or nominally thirteen) Territorial divisions. The appointment in 1928 of a Director of Mechanization promised speedier progress. Nevertheless ten years elapsed before the Regular Army had its full complement of wheeled vehicles and half its complement of tracked vehicles other than tanks.[10]

Re-equipment with modern weapons was, for many reasons, a more difficult task. When the government rejected the Defence Requirements Committee's proposal that the field force should be made ready for prompt despatch to the Continent, the standard field gun in the British Army was an 18-pounder dating from World War I and mounted on a carriage whose wheels were without pneumatic tyres. This could be brought up to date by relining and the provision of a better carriage, but these improvements had yet to be made although the matter had been under discussion for many years. The standard anti-aircraft weapon in service in 1935 was a 3-inch gun also dating from World War I and with an effective ceiling of only 17,500 feet. The army's automatic weapons were the Lewis gun, designed in 1912, and the Vickers machine-gun developed at Erith before the end of the nineteenth century but afterwards improved. In 1936 the army possessed only 375 tanks, of which 304 were officially classed as obsolete.[11] All but two of its 166 medium tanks were out of date and only 69 of its 209 light tanks were modern. These were Light Tanks Marks V and VI, designed by Sir John Carden for Vickers-Armstrongs and put into production about the time when Carden's death in an air crash deprived the nation of the one man whose talents as a designer of armoured fighting vehicles could be said to approach genius.

One of many factors which made the rearming of the disarmed army a daunting task was a severe lack of productive capacity. In 1918 the Royal Arsenal at Woolwich had employed about 65,000 people, the Royal Small Arms Factory at Enfield Lock about 9,500 and the Royal Gunpowder Factory at Waltham Abbey about 5,700. By the early 1930s these numbers had shrunk to 7,000, 800 and 350 respectively. Only one of the national munitions factories set up during the war had survived, and that was on a care-and-maintenance basis. Before the 1914–18 war at least four large and some twelve smaller commercial firms had habitually tendered for War Office contracts for the supply of guns, small arms and ammunition. In 1935 there were no fully-fledged armament firms except Vickers-Armstrongs. The Inter-service Principal Supply Officers' Committee maintained a list of contractors whose services might be drawn upon in time of war, but very few of them had any immediate capacity for the production of weapons. William Beardmore and Company had some gunmaking capacity but, insofar as they were still arms manufacturers, were chiefly concerned with naval requirements. So indeed were Vickers, but they also had a considerable capacity for the production of army weapons. Armstrong Whitworth Securities still owned the works at Scotswood used during the war as a munitions factory, and in 1937 they sold it to the Admiralty, who leased it to Vickers. The Birmingham Small Arms Company, better known as BSA, were experienced manufacturers of small arms; Greenwood and Batley Limited and Imperial Chemical Industries had a limited capacity for the manufacture of small arms ammunition. Imperial Chemical Industries also had some capacity for the production of military explosives, but this would remain fairly small as long as rearmament was not allowed to interfere with normal trade. Firms established during the rearmament era for the sole purpose of manufacturing war material included the British Manufacture and Research Company, New Crown Forgings Limited and Nuffield Mechanizations and Aero Limited. In Canada a shipbuilding and salvage firm, Marine Industries Limited, accepted an order for a hundred 25-pounder field guns, but the contract was signed only just before the outbreak of war.

Capacity for design was limited to an even smaller number of firms or agencies. In the private sector it was virtually confined, when rearmament began, to Vickers-Armstrongs and (for small arms only) BSA. In the public sector, there existed at Woolwich a Design Department—formerly the Drawing Office—whose primary duty was defined as the preparation for any of the three fighting services of original designs of guns, small arms, ammunition, bombs, pyrotechnics and certain other warlike stores—but not tanks. Originally the Design Department merely prepared drawings in accordance with detailed specifications issued by the Ordnance Committee, but eventually the disappearance of most commercial firms with a capacity for design forced it to assume a more active role. By the middle of the 1930s the usual procedure when a new weapon was wanted for the army was that the War Office made its requirements known in general terms and the Ordnance Committee prepared a detailed specification as before. Vickers-Armstrongs and the Design Department were then invited to produce definitive designs and prototype or 'pilot' weapons which were subjected to comparative tests. Eventually one of these was approved and either an order was placed or the project was shelved until funds for production of the weapon were available. Alternatively, only Vickers-Armstrongs or only the Design Department might receive the specification and be asked to develop the weapon on a non-competitive basis.

Partly because the Ordnance Committee was composed chiefly of retired or very senior naval and army officers with a natural bias towards orthodoxy, this procedure tended to yield rather conventional weapons whose derivation from their forerunners was unmistakable. The system did not debar a commercial firm which could afford to do so from developing a weapon as a private venture and offering it to the government, but such ventures were risky because the authorities seldom accepted a design which called for scarce or novel materials or which might, in their opinion, lead to production difficulties if the weapon had to be manufactured in wartime by firms outside the industry. The Vickers-Berthier machine-gun was a case in point. In 1918 Vickers, foreseeing an eventual demand for a new light machine-gun, acquired an interest in such a gun, the Berthier. They spent more than £100,000 on patent rights and development, only to see the Vickers-Berthier turned down in favour of the Zbrojowka 2B, first made at Brno in Czechoslovakia. A version of the Zbrojowka 2B modified by the Design Department was accepted for production in 1935, four years after the specification was issued and some ten years after the adoption of a new light machine-gun was first mooted. Deliveries of the Enfield-made version, called the Bren, began in 1937, and in 1938 five thousand Brens were ordered from a Canadian firm, the John Inglis Company. One of the reasons given for rejection of the Vickers-Berthier was that it was difficult to produce, but it was bought by the Indian government and manufactured in India by 'raw native labour'.

Such delays were not uncommon. The 25-pounder field gun used by the British Army throughout the Second World War went into production some thirteen years after replacement of the 18-pounder and the old 4·5-inch howitzer by a single weapon was first proposed. A 3·7-inch successor to the 3-inch anti-aircraft gun was under discussion by 1920, but the specification was not issued until 1933 and production did not begin until 1938. The No. 4 Rifle and the Boys Anti-Tank Rifle were first produced some sixteen and nine years respectively after they were first mooted.

However, it cannot be assumed that quicker methods would have yielded better results. Time had to be allowed for research and development, for modifications to be made in the light of practical trials and sometimes of reports about developments in other countries. The system was intended to provide the army with weapons which would stand up to rough treatment and were unlikely to be superseded for many years to come. On the whole these aims were achieved so far as artillery and small arms were concerned, but it sometimes happened that the search for a new weapon began too late. For example, in 1936 an urgent demand arose for a weapon capable of dealing effectively with low-flying aircraft. No such weapon was ready, chiefly because the authorities had cherished until recently the illusion that some form of machine-gun would suffice. Vickers had prepared drawings of a 40-millimetre 2-pounder which might be adapted for the purpose, but it was transportable rather than fully mobile and had other disadvantages. Time was short, so the existing 40-millimetre Bofors was tried. It lived up to claims made for it by the Swedish Minister in London, and it had the additional merit of working well with British ancillary equipment. An order for Swedish-made guns was placed, and in 1937 Nuffield Mechanizations and Aero Limited and a new Royal Ordnance Factory specially equipped for the production of anti-aircraft weapons began to manufacture the weapon under licence.

The Design Department had no counterpart in Germany. The A-4 or V2 rocket was designed and developed by an offshoot of the ballistics and munitions section of the German Army's ordnance department with some help from Siemens and later from academic scientists; but this was an exceptional case, where secrecy was paramount. In general, the Germans relied for the design and development of their weapons on commercial firms in which the state might or might not have a financial interest. Even the FZG 76 pilotless aircraft or V1, although essentially a secret weapon like the A-4 rocket, was in all important respects the creation of Argus Motorenwerke, Gerhard Fieseler and Askania, all commercial firms. The German Air Ministry appointed a technical officer to co-ordinate the work of the three firms but contributed very little to the design. As a rule the authorities made no stipulations about ease of manufacture but gave designers a free hand to meet official requirements in their own way.

Differences between the British and German systems do not seem to have had any significant effect on the relative merits of the artillery and small arms used by one side and the other in World War II. Armoured fighting vehicles were another matter. Even so, the enormous difficulty experienced by the British in providing themselves with tanks that enabled them to win battles cannot be attributed to a single cause. It was bound up with such fundamental issues as the army's role in war, the terrain over which the war would be fought and the kind of contribution tanks would be called upon to make.

The government's rejection in 1934 of the Defence Requirements Committee's proposals for the prompt despatch of the field force to the Continent in the event of war left these issues unresolved. The Secretary of State for War regarded the government as still committed to the eventual despatch of an expeditionary force which would have to be reinforced by the Territorial Army, but his colleagues were not easily persuaded that the land forces should make more than a token contribution to any future European war. A review of the roles of all three fighting services led the Minister for the Co-Ordination of Defence to conclude in 1937 that their most important task was the air defence of the United Kingdom. The security of the trade routes must come next, and after that the defence of the overseas territories of the Commonwealth. This left co-operation with any European allies Britain might have in fourth place. The principal task of the Regular Army was defined in the light of this list of priorities as 'the defence of imperial commitments, including anti-aircraft defence at home'.[12] The Territorial Army was to perform 'duties in connection with the maintenance of order and of essential services'.[13]

The Committee of Imperial Defence confirmed early in the following year that procurement programmes should be based on the assumption that only a limited contribution would be made to a European war. The five Regular divisions of the field force—which became six when the mobile or cavalry division split into two divisions—were to be equipped on a scale suitable for 'colonial warfare in operations in an Eastern theatre'. Since the Cabinet had already ruled that the Territorial Army should be trained to use the same weapons as the Regular Army, it followed that Territorial divisions other than those assigned to the air defence of Great Britain would also be equipped on a colonial scale.

In theory, the making-good of the army's worst deficiencies continued to be governed by these principles until the spring of 1939. The government, hitherto unwilling to sanction staff talks with the French at a higher level than that of the service attachés, then jettisoned the principle of limited liability by agreeing to talks at the highest planning level. These opened about the time when the Germans began to move into Bohemia and Moravia. The British undertook to send across the Channel on the outbreak of war four divisions as the first instalment of a force of thirty-two (afterwards increased to thirty-six). They proposed to find the thirty-two divisions by doubling the Territorial Army (nominally of thirteen but in fact of twelve divisions) and adding the twenty-six divisions thus obtained to the six Regular divisions of the field force. Ultimate expansion to a total of fifty-five divisions was envisaged but not formally approved until the spring of 1940.

In practice the military authorities did not allow such a vague definition as 'operations in an Eastern theatre' to prevent them from doing what they could to provide the army with the armoured fighting vehicles it would need if it had to send an expeditionary force across the Channel and also to meet commitments which included a contingent liability for the defence of Egyptian territory against an Italian invasion from Libya.[14] Their task was made extremely difficult by the knowledge that, although no one could predict the course of a European war, it would almost certainly develop along different lines from a campaign fought in the Western Desert of Egypt.

When rearmament began there was no capacity for the design of tanks outside Vickers-Armstrongs. The Design Department's tank experts had moved in 1930 to the Mechanization Board, which did not then design tanks but merely passed judgment on designs submitted to it and suggested modifications and improvements. Until 1936 the weight of a tank was limited by a League of Nations convention to a maximum of sixteen tons, and engineer units of the British Army were not equipped to build temporary bridges capable of bearing even as big a load as that. Designers were also hampered by a limited choice of engines and by the stipulation that a tank must be narrow enough to be transportable by rail without interference with traffic on adjacent lines. Furthermore, for many years there was a tendency in British military circles to underestimate both the tactical value of tanks and their mechanical complexity. The steering and suspension of a medium or heavy tank involved problems of design which could not be tackled at short notice by general engineering firms with no previous experience of large tracked vehicles.

Although the tank was a British invention the British tended, too, to misunderstand and misinterpret its rôle in modern warfare. In World War I slow, heavy tanks had been used to break through entrenched positions, but even then the term 'landship' was sometimes used. After the war false analogies with naval warfare led not merely to the application to tanks of such inappropriate terms as 'cruiser' and 'destroyer' but to the concept of tank battles in which armoured fighting vehicles would be used not so much to gain a positive tactical advantage as to seek out and destroy the enemy's armour. Tanks came to be regarded primarily as tank-destroyers, destruction of the enemy's tanks as an end in itself.

These factors combined to bring about a situation in which, as the Director-General of Munitions Production admitted, the real difficulty where tanks were concerned was to know what was wanted. In 1936 the War Office came to the tentative conclusion that the cavalry needed a light tank for reconnaissance and skirmishing, the Tank Brigade—at that time the only armoured brigade—a more powerful 'cruiser' tank for armed reconnaissance and a medium tank to provide hitting-power. In addition, a thickly-armoured tank would be needed for infantry support.[15] But light tanks were soon to seem less desirable in the eyes of cavalry commanders than 'cruiser' tanks and armoured cars.

The only satisfactory British tanks of recent design in service or about to come into service in 1936 were the Light Tank Mark V designed by Sir John Carden for Vickers-Armstrongs shortly before his death and the Light Tank Mark VI developed from it. Carden had, however, also designed two tanks which became, with modifications made after his death, the Cruiser Tank Mark I and the Cruiser Tank Mark II, and had begun work on a tank with 60-millimetre armour which became Infantry Tank Mark I and was the first of two tanks popularly called Matilda. The second Matilda was Infantry Tank Mark II, developed from Infantry Tank Mark I by the Mechanization Board and Vulcan Foundry.

The pilot models of these four tanks were numbered by the War Office A9, A10, A11, and A12. Soon after the first two made their appearance, interest in 'cruiser' tanks was stimulated by reports from British observers of promising results obtained by the Russians at manoeuvres with tanks comparable with the A9 and A10 but described as faster and likely to be easier to produce. The Russian tanks were based on the American Christie chassis, which Oliver Bowden of the Nuffield organization had already brought to the attention of the War Office. Trials of the Christie chassis in the winter of 1936–37 showed that radical changes would be needed before the chassis of an American ten-ton tank could be used as the basis of a twelve-ton British one. Development of the Christie-based A13 Marks I and II by Nuffield Mechanizations and Aero Limited proved a difficult business, and the Cruiser Tanks Marks III and IV eventually produced were not a great success. Cruiser Tank Mark V (the Covenanter) was a modified version of Mark III with a hull designed by the London, Midland and Scottish Railway, a Meadows engine and a turret designed by Nuffield Mechanizations. It proved disappointing and was never used in action as a combat vehicle.

Of these seven tanks (numbered from A9 to A13 Mark III) the best were the three designed wholly or in part by Carden, and the A12, officially credited to the Mechanization Board and Vulcan Foundry but derived from the A11. After accepting the A12 for production as Infantry Tank Mark II, the authorities invited Vickers to manufacture it, or alternatively an infantry tank derived from their own A10. The outcome was the Infantry Tank Mark III, called by Vickers the Valentine because it was submitted to the War Office on or about St Valentine's Day in 1939.

Not surprisingly, Britain's output of tanks before the war was very much smaller than that of the Germans and the French. Early in 1937 production at Elswick was transferred to a new workshop on the site of the old shipyard. In the course of the year the War Office ordered a hundred light tanks, fifty Mark I cruisers and sixty Mark I Matildas. No Mark I cruisers were delivered before 1939, but the output of light tanks reached twenty a month in the early part of 1938, and the first sixteen Matildas were delivered in the autumn of that year. Meanwhile the A10, intended by Vickers as an infantry tank, was relegated to limbo, only to be revived as a cruiser in 1939 and put into production at short notice. Altogether about 720 light and 130 cruiser or infantry tanks reached the army between the Munich crisis and the outbreak of war.[16] The output of light tanks climbed from 42 a month in the last quarter of 1938 to 91 a month in the second quarter of 1939, but fell to 34 a month in the last two months of peace. Deliveries of cruiser and infantry tanks remained almost stationary at six to eight a month until the late summer of 1939, when they soared dramatically.

*

On the eve of World War II the United States, like Britain, possessed a numerically strong navy but only a small army. The active strength of the United States Army in September 1939 was approximately 190,000 of all ranks.[17] The Officers' Reserve Corps numbered about 110,000. There was also a partially trained National Guard about 200,000 strong. Nominally there were four field armies in the Continental United States, but they had no staffs and only four organized and seven partly-organized divisions under command. The United States Army Air Corps had about 1,800 aircraft, mostly obsolete or obsolescent.[18] The army's standard weapons were the Springfield rifle, a 75-millimetre field gun and a 3-inch anti-aircraft gun, all dating from the First World War. A new Garand rifle had been chosen, but production was only about 4,000 a month. Few, if any, of the army's 329 tanks were up to the demands of modern war, and deliveries of new light tanks did not exceed thirty a month. Stored reserves of older weapons included more than 2,500,000 rifles, 113,000 machine-guns and nearly 9,000 field guns, but there were not enough modern weapons in the whole country to equip even one division. There was virtually no current production of such weapons outside the seven national arsenals.

The *potential* capacity of American industry, on the other hand, was impressive. How it should be exploited if the need arose had been the subject of a good deal of study. By the terms of the National Defense Act of 1920, the Assistant Secretary for War was charged with the procurement of all military supplies and with forward planning, but the General Staff were responsible for mobilizing manpower and material resources in the event of war. In practice, the formulation of military requirements and the distribution of equipment and supplies were undertaken by the G-4 or Supply Division of the General Staff. The Assistant Secretary of War, Louis Johnson, announced on 30 June 1939 that his emissaries had inspected 20,000 plants and that 10,000 firms had been selected as wartime suppliers of military equipment and furnished with production schedules which would become effective on mobilization. Only a few of these, of course, were potential suppliers of weapons as distinct from uniforms, non-military hardware and general stores. The Assistant Secretary also said that in wartime the Army Air Corps would need five or six thousand aircraft a year. This was about three times the current output.

But the Assistant Secretary of War and the General Staff were concerned only with the army's needs. Questions that had to be considered at a higher level were how industry should be organized in the light of imminent developments in Europe, and how its output should be apportioned between the armed forces, civilian consumers, and foreign governments whose attempts to halt the march of totalitarianism the United States government might think it expedient in the national interest to support.

The second of these questions was made difficult by recent legislation which forbade the export of arms and aircraft and the granting of loans to belligerent powers. It was further complicated by a ban imposed by the Johnson Act of 1934 on the granting of loans, except renewals or refundings, to any government either wholly or partly in default with respect to its financial obligations to the United States government. The British were popularly supposed in the United States to have omitted to pay for large quantities of American goods bought during World War I, but that was not quite accurate. Between 1914 and 1919 they had bought for their own use or on behalf of their allies American war material to the value of approximately twelve thousand million dollars. All American suppliers had been paid in full. The British had found the money by remitting gold or selling foreign securities to the value of eight thousand million dollars and borrowing the remaining four thousand million dollars from the United States government. Towards the four thousand million dollars they had paid two thousand million by way of principal and interest up to the time in 1933 when the world economic crisis led to a general suspension of such payments. The fact remained that, irrespective of the provisions of the Johnson Act, as long as the Neutrality Act of 1937 remained in force neither the British nor the French would be able if they went to war to buy American arms or aircraft, or even to take delivery of arms or aircraft already ordered and paid for. Nor would their governments be able to pay even for non-military materials by raising loans or credits in the United States. Commercial firms would, in theory, be able to do so, but whether in practice even that would be allowed was doubtful.

At the end of 1938 President Roosevelt and the Secretary of State, Cordell Hull, came to the conclusion that the Neutrality Act was an incitement to war, inasmuch as it gave the totalitarian powers an inducement to force Britain and France into belligerency and thus deprive them of American supplies.[19] Roosevelt made up his mind that such a state of affairs must not be allowed to continue, but was determined not to repeat the mistake made by President Wilson when he laid his policy in ruins by failing to carry the Senate with him. A Bill to amend the Neutrality Act was presented to the Senate in the spring of 1939, but without success. The outcome of further attempts to obtain a sympathetic hearing in the Senate and the House was that on 12 July the Senate Foreign Relations Committee decided by a majority of one to shelve the Bill until January 1940. The President and Hull, fortified by a stream of telegrams from foreign capitals which predicted imminent war and described the failure of the United States to lift the arms embargo as an encouragement to the dictatorships and a discouragement to the democracies, then put their case to the party leaders in Congress, but were assured by Senator William E. Borah on 18 July that there would be no war in the near future.

There the matter rested until, on 1 September, the Germans invaded Poland. Thereupon the administration decided that a special session of Congress should be summoned for the purpose of revising the Neutrality Act. The date fixed was 21 September.

ELEVEN

World War II: The Lull and the Storm

Hitler's occupation of Bohemia and Moravia in the spring of 1939 put the Poles in a highly vulnerable position. In the event of war with Germany, German troops with strong air support would be able to threaten them not only from the north and east, but also from the south. The Poles, with an army of thirty Regular and ten reserve divisions and an air force of some five hundred first-line aircraft, would have to choose between a suicidal attempt to hold their forward positions and a strategic withdrawal which would leave their centres of production in Silesia uncovered. Once these were lost, they would depend for the means of continuing the war on such supplies as might reach them across their frontiers with the Soviet Union and Rumania.

Within seventy-two hours of Hitler's arrival in Prague, the British sounded the Russians about their attitude to the Polish question. The Russians accepted in principle their proposal that Britain, France, Poland, Rumania and the Soviet Union should make a joint declaration to the effect that any further act of aggression by Germany would be resisted, but suggested a preliminary conference at Bucharest to which the Turks should be invited to send an emissary. But the Poles were reluctant to enter into any arrangement that might give the Russians a pretext to send troops into Poland, although they still hoped to receive supplies and possibly air support from the Soviet Union. Against the background of Hitler's seizure of Memel in Lithuania, they asked for explicit promises of British and French support. Neville Chamberlain, after hurried consultation with the French, then made his fateful announcement that Britain and France would resist any act that clearly threatened Poland's independence. Later he threw in Rumania and Greece for good measure.

But the British, as the veteran Lloyd George reminded

8*

Chamberlain in the House of Commons, would not be in a position to give any effective help to the Poles unless and until they succeeded in coming to terms with Russia. Their chances of doing so suffered a sharp setback when, in May, the pro-Western Maxim Litvinov was replaced as Soviet Commissar for Foreign Affairs by the inscrutable Vyacheslav Molotov. The signing of the Moscow Pact on 23 August came as a great shock to the British public. But it did not surprise the French Ambassador in Berlin, who had predicted a few days after Litvinov's dismissal that Hitler would strike a bargain with Stalin.

The economic and strategic consequences of the Moscow Pact were even more far-reaching than they seemed at the time to observers too deeply shocked by its iniquity to grasp all its implications. By depriving the Poles of any chance of maintaining their forces in the field for more than a month or two if they retreated eastwards, the pact made the defeat of Poland well-nigh certain even if the Russians remained on their own side of the frontier. By giving Germany access to rich sources of food and raw materials in Eastern Europe, it still further reduced Britain's already slender chances of making serious inroads on her industrial capacity by economic blockade.

So the ultimate consequence of Chamberlain's gesture of defiance was that, so far as the fate of Poland was concerned, Britain and France went to war on 3 September for a cause already lost. A vigorous offensive against the weakly-held Siegfried Line might conceivably have made such an impression on the German High Command that officers of the Wehrmacht who had been toying with the idea of getting rid of Hitler would have steeled themselves to make the attempt, but whether it would have been successful is another matter. In any case the Allies had no intention of going to such lengths. The aims they set themselves on the outbreak of war were to parry as best they could the expected knock-out blow from the air, give some indirect assistance to the Poles by pinning a number of German divisions to the Western Front, and build up their strength in the hope that sooner or later the enemy would exhaust himself by attacking their supposedly impregnable positions. Various proposals for limited offensives against Germany or her sources of supply were discussed during the winter of 1939–40, but none of them had come to anything when, in April, the Germans took the initiative by striking at Denmark and Norway.

Not surprisingly, the mobilization of the French Army and the despatch of the first wave of the British Expeditionary Force across the Channel seemed to Hitler little more than an elaborate charade. Until events proved him wrong, he was confident that, once the conquest of Poland was an accomplished fact, the British and the French would be glad to make peace on terms which left him with a free hand in Eastern Europe. When they declined to do so, he insisted that an offensive should be launched on the Western Front without delay. Much to the relief of his service chiefs, the weather provided them with valid reasons for cancelling or postponing attacks planned for various dates between November and January.

In all the belligerent countries, mobilization temporarily reduced industrial output or retarded its growth. In Britain responsibility for equipping the army had passed about a month before the outbreak of war to a newly-formed Ministry of Supply, staffed largely by recruits from the production side of industry. Deliveries of new 25-pounder field guns and 3·7-inch anti-aircraft guns increased scarcely at all during the first six or seven months of the war.[1] Deliveries of tanks improved from an average of 58 a month in the last two months of peace to nearly 80 a month in the first four months of war, but the proportion of cruiser and infantry tanks was still unsatisfactory. Deliveries of aircraft, on the other hand, after falling from an average of 775 a month between September and November to 600 in December, rose smartly in 1940. In France, 135,000 key men who should have stayed in their peacetime jobs were called up because their applications for exemption had been made too late. The Germans produced between September 1939 and April 1940 more than half a million rifles, some twenty-seven thousand machine-guns and nearly fifteen hundred field or medium guns, but only about five hundred tanks. They formed more than forty new infantry divisions by May, but were able to equip them and to re-organize and expand their armoured forces only by drawing on stocks as well as current output. German aircraft production climbed slowly from a monthly average of just under 700 in 1939 to just over 900 in 1940, as compared with British monthly averages of 660 and 1,250 respectively.

In the light of these figures it does not seem surprising that the Allies felt in the winter of 1939–40 that time was on their side. When September and October passed without bringing the dreaded air attacks on their centres of population and production, they were encouraged to believe that they were bound to win in the end. As Chamberlain put it, the only question was how long it would take them to achieve their purpose.

In November the Americans duly lifted their embargo on the sale of arms to belligerents, but not their ban on loans and credits. The British, estimating that their gold reserves and overseas investments exchangeable for dollars might last up to three years with careful management, did not contemplate the placing of large orders for American munitions unlikely to be available in substantial quantities for eighteen months or more. They preferred to regard the arms-making potential of the United States as a reserve to be drawn upon in an emergency. They did, however, need machine-tools, chemicals and explosives, and they had already placed with Lockheed and the North American Aircraft Company orders for Hudson bomber-reconnaissance aircraft and Harvard trainers which were large by American standards. They had also ordered more than four million dollars' worth of anti-aircraft predictors from the Sperry Gyroscope Company, to say nothing of five-and-a-half million feet of rubber hose which did not count as war material but was to prove invaluable when London and provincial cities were under attack in 1940 and 1941. Apart from these, they expected to need large quantities of copper, molybdenum, raw cotton, and steel forgings and castings. With the approval and active help of the United States Treasury, they and the French set up purchasing commissions which worked in close co-operation under the general direction of an Anglo-French Co-Ordinating Committee in London.

As things turned out, the careful husbanding of Allied gold and dollar resources proved far from easy. The amended neutrality regulations barred American merchant vessels from British, French, German, Belgian, Danish, Dutch, Irish and Swedish ports, from the whole of the Baltic and from Norwegian ports south of Bergen. American goods bought by the British and French governments had not only to be paid for on the spot, they had also to be fetched. To provide part of the necessary carrying capacity, the British spent about eighteen million dollars by the summer of 1940 on the purchase of used cargo vessels from American shipping firms. Later in the year the War Cabinet sanctioned the placing of orders for sixty new ships at an estimated cost of eighty million dollars. Contracts between the British government and the Todd-Bath and Todd-California Shipbuilding Corporation entailed the construction of two brand-new shipyards and launched the United States on a massive shipbuilding programme which raised the output of American shipyards from little more than half a million tons in 1940 to nineteen million in 1943.

Dollars had also to be found for the construction of new factories by firms willing to accept orders from the British government but not to meet the cost of consequent additions to their plant. Fourteen transactions of this kind were negotiated in 1940 at a capital cost to the British of some thirty-seven million dollars. By far the most important were those which arose from approaches made to E. I. du Pont de Nemours and Company and the Hercules Powder Company. Two new companies, wholly owned by the British government but incorporated under American law, were set up in Tennessee and New Jersey. Du Pont and the Hercules Powder Company contracted to build new explosives factories with funds remitted to them through these companies, and to sell their output to the British government. Eventually the three factories built in accordance with these arrangements were taken over by the United States government at approximately their cost price of some thirty-five million dollars, but their output up to various dates in 1942 was paid for in cash and cost the British about fifty million dollars.

These were exceptional measures, adopted to tide over an emergency. But in war emergencies tend not to be exceptional. If the British needed shipping and explosives, the French needed aircraft. Their statesmen were so painfully conscious of the shortcomings of the Armée de l'Air that the time was near when a French Prime Minister would declare himself willing to sell every picture in the Louvre if the proceeds would buy American aircraft. The limiting factor was not, however, money but productive capacity. By the end of 1939 orders placed by the French in the United States since 1938 totalled about 2,000 aircraft, and the Pratt and Whitney Aircraft Company had already doubled its capacity for the purpose of meeting an order for 6,000 aero-engines. Orders placed by the British since the outbreak of war would amount by the following February to 1,320 aircraft (about half of them trainers for Canada and New Zealand), and they had ordered 1,200 engines. Virtually the whole of the existing capacity of the American aircraft industry was absorbed by work in hand for British, French and American buyers, so any increase would have to come from additions to plant and floor-space.

At that stage British and French representatives in Washington discussed with President Roosevelt a tentative plan for the purchase on joint British and French account of a further 10,000 aircraft and 20,000 engines. The British government, alive to the risk that large contracts might tempt American manufacturers to order so many machine-tools that few would be available for export, was not enthusiastic about the scheme, but agreed that a British emissary should accompany a French mission about to leave for the United States.

The mission reported early in 1940 that the American aircraft industry would undertake to deliver 8,400 aircraft between October 1940 and September 1941.[2] These would be additional to aircraft already ordered. The cost was estimated at fifteen hundred million dollars, and this figure would include upwards of five hundred million dollars for capital assistance to firms unable or unwilling to increase their productive capacity at short notice without such help.

British experts did not believe that the Americans would be able to produce anything like 8,400 aircraft in twelve months. Nevertheless the Air Ministry recommended that, with certain safeguards, the scheme should be accepted. Even if little more than half the aircraft promised were completed, the scheme would be a valuable insurance against the destruction of Allied factories by bombing. It would enable the Allies to achieve air superiority somewhat earlier than they could otherwise hope to do, and would show the world that American industry was behind them.

Eventually, in the light of further discussion and a second visit to the United States, the target figure was reduced to 4,600 aircraft, with spares. Altogether 2,440 fighters and 2,160 bombers of ten different types, with 8,000 engines of five different types, were to be manufactured on joint British and French account. The Supreme War Council approved the scheme in March 1940 on the understanding that the British were to receive just over half the aircraft completed and that the French were free to take up an option on an additional 4,500 engines.

The French were not destined to derive any benefit from these arrangements. Indeed, they were unable to take delivery of more than about half the aircraft ordered earlier. As things turned out, the most important consequence of the scheme was that it gave the American aircraft industry a boost which made all the difference to the subsequent expansion of the United States air forces.

Meanwhile the Americans were taking their first cautious steps towards rearmament. During the first nine months of the war in Europe almost imperceptible additions were made to the Regular Army and the authorized establishment of the National Guard. Officers were called from the reserve in small numbers for short tours of active duty. Warships made 'neutrality patrols' off the Atlantic seaboard, but the main fleet was stationed in the Pacific. The Joint Army and Navy Board instructed its planners to prepare outline plans for a war fought alone, or in partnership with the British and the French, against Germany, Italy, Japan, or any combination of those powers. The framing of these plans called attention to the possibility that eventually land and air forces might have to be sent to Europe or Africa, or to both, to join the British and the French in fighting one or more of the Axis powers, but in 1939 that possibility seemed remote. The emphasis was on the defence of the Americas and the safeguarding of American interests and possessions in the South-West Pacific.

Meanwhile the most important contributions to national preparedness were those made not by the administration or the armed forces but by foreign governments. Orders placed by the British and the French for aircraft, ships and explosives did far more to pave the way for rearmament than anything the President or the Chiefs of Staff could have done before the spring of 1940. 'Without the head-start given by these foreign orders,' said the Secretary of War in 1941, 'we should at the present time be in a very grave situation.' The Secretary of the Treasury put the length of the 'head-start' at eighteen months.[3]

*

In the late spring and early summer of 1940 the Germans defeated the French Army in five weeks. The British Expeditionary Force escaped with the loss of most of its tanks, guns and transport.

Spokesmen for the Vichy government and some of the defeated generals attributed the fall of France to moral decay and a massive inferiority in weapons and equipment. Marshal Pétain said that the nation was undone by 'laxity and pleasure-seeking'. The implication was that the traditionally frugal and industrious French had become so effete that they failed to arm themselves.

It would seem only commonsense to ask, before accepting that verdict, what resources the French actually had when the Germans opened their offensive.

So far as warships and naval weapons were concerned, their resources were more than adequate. Although the *Jean Bart* and the *Richelieu* had yet to be completed, the French Navy was far stronger in almost every respect than its German counterpart.

Nor were the French short of troops. The Germans had available for their campaign in France and the Low Countries eighty-nine divisions in Army Groups A, B and C and forty-five in OKH Reserve. The French had on their North-East Front or in rear of it ninety-four of their own and ten British divisions. The Belgians had some twenty-two divisions, the Dutch about eight. So the number of major formations on each side was about the same.

As for artillery, the French had a numerical superiority of some thirty-five per cent in field and medium guns, and their heavy guns outnumbered the 280-millimetre mortars on which the Germans relied by more than five to one. To see what kind of artillery support they were capable of providing in a particular case, let us take the sector at Sedan whose capture by the German 19th Panzer Corps on 13 May ranks as a classic example of the successful use of *Blitzkrieg* tactics.

This was regarded by the French High Command before the battle as a 'safe' sector. The French 55th Division, standing on a five-mile front, had only to prevent the enemy from crossing an unfordable river sixty yards wide and establishing himself on the left bank. The road-bridge across the Meuse at Gaulier, in the outskirts of Sedan, was blown about 7 p.m. on 12 May, so the 19th Panzer Corps would not be able to bring tanks or guns across the river until it held both banks and could either repair the Gaulier bridge or erect a temporary structure.

The main line of defence consisted of trenches and concrete pill-boxes. The pill-boxes were spaced at intervals of approximately two hundred yards, and each was armed with an anti-tank gun and machine-guns. Although the 55th Division did not have its full complement of anti-tank guns, there were also anti-tank guns on its left flank, where the river made a loop. A second line of defence, at the foot of the slope behind the Meuse, consisted of trenches and barbed wire, without pill-boxes.

The 55th Division, like most of the French Army, was short of anti-aircraft guns. Medium and light machine-guns on the hillside in rear of its second line of defence were, however, well placed to prevent hostile aircraft from diving into the valley. Moreover, when German aircraft attacked a French mechanized cavalry column and other objectives on the right bank of the Meuse on 12 May, they were soon driven off by Curtiss Hawk fighters which claimed the destruction of thirty aircraft without loss to themselves.

Artillery support for the troops in this supposedly secure sector was provided by roughly twice the normal divisional share of field guns, supplemented by medium and heavy guns from corps and army reserves. No doubt this fairly generous allotment —about 140 guns in all[4]—was at least partly attributable to the excellence of the artillery positions available. Observation posts on the heights above the Meuse commanded the ground on the right bank up to the edge of the forest of the Ardennes, about six miles away.

So the 55th Division was not defeated on 13 May for lack of guns. The chief cause of the disaster which overwhelmed it on that day was a series of dive-bombing attacks on its positions. These were spread over the best part of five hours. They did little damage and caused few casualties, but some of the defenders were so dazed and cowed by a long period of almost continuous bombing that they failed to make good use of their weapons. Infantry, riflemen and dismounted motor-cyclists of the 19th Panzer Corps were able to cross the river in collapsible boats and establish themselves on the left bank. The fighters which had intervened so successfully on the previous day did not appear.

Was this because the Germans, as is commonly supposed,

were overwhelmingly superior in the air? That is not an easy question to answer, because other factors besides numbers have to be taken into account. Certainly the Luftwaffe had a big advantage in striking power. To support their troops in France, Belgium and Holland, the Germans could call on some 300 to 400 dive-bombers and approximstely 1,300 long-range bombers. Eight months after the outbreak of war, the Armée de l'Air still had only about 400 bombers of all classes. Fewer than half the American medium bombers ordered in 1939 reached France by the summer of 1940, and the output of French factories was disappointing. The air formation serving the First Army Group, on the North-East Front, had only seventy-four bombers or ground-attack aircraft of its own. The British contributed eight Battle and four Blenheim squadrons stationed in France and four Blenheim and two Whitley squadrons based in England—a total of some 200 to 220 aircraft. All these aircraft, British and French, proved highly vulnerable to anti-aircraft fire. Unless heavily escorted, they were also an easy prey for fighters. Once the Germans held both banks of the Meuse at Sedan and elsewhere, the French were unable to prevent them from passing tanks and guns across the river, because they lacked the means of mounting effective bombing attacks on repaired or newly-installed bridges.

However, they did not need bombers to prevent the Germans from ever reaching the left bank. Their infantry and artillery in the threatened sectors ought to have been able to do that. They failed to do it partly because their training and tactical doctrine were defective, partly because the fighter support needed to give them confidence was lacking.

How far the absence of fighter support was due to a shortage of aircraft is an interesting question. When the Germans opened their offensive on 10 May, they had some 860 single-seater fighters ready for service on the Western Front. (They also had some 350 heavy fighters, but these proved more successful as ground-attack aircraft than in the long-range escort role for which they were intended.) The Armée de l'Air had 680 single-seater fighters in metropolitan France on that day, and an average of 153 fighters a week for the next five weeks should have put it in a fairly good position to replace losses. The British contribution rose from roughly 70 aircraft on 10 May to some 200 a week later. Unless the French suffered far heavier losses between

10 May and 12 May than those admitted in published accounts, they can scarcely have been desperately short of fighters on the fourth day of the battle.

They were, however, handicapped by an inflexible distribution of their forces. The Germans had evolved a system by which air support could be switched without delay to the part of their front where it seemed to be most needed. Under the French system, air units were assigned in advance to particular tasks. The air formation responsible for supporting the First Army Group was allotted about half the Armée de l'Air's twenty-four single-seater fighter wings, but five of these wings were hypothecated to the defence of Paris and the Lower Seine. For practical purposes this meant that roughly 200 French fighters, supplemented by such British fighters as might be available, had to cater for all the needs of the French Seventh Army on the extreme left, the British Expeditionary Force on the Seventh Army's right, the French First Army on the British right, and the French Ninth and Second Armies on the Meuse. In the light of this figure one learns without astonishment that the Curtiss Hawk fighters which intervened at Sedan on 12 May were unable to help the 55th Division on 13 May for the simple reason that they were sent elsewhere. Dewoitine 520 fighters which could also have been sent to Sedan on 13 May were absent for the same reason.

It seems fair to conclude that, while admittedly the French were seriously short of bombers, their failure to provide adequate fighter support for their troops was due at least as much to an unsound organization as to any lack of numbers. French aircraft manufacturers were—to say the least—not solely to blame for the industrial and political upheavals which hampered their output of bombers in the 1930s. They were in no way responsible for the poor use made by the authorities of the fighters they did produce.

In regard to armoured fighting vehicles, we have already seen that Renault, Hotchkiss, Somua, and the Forges et Chantiers de la Méditerranée produced under the four-year plan of 1936 some 3,400 good tanks of recent design. In the spring of 1940 about three-quarters of these were held by first-line units available for service on the North-East Front. The Germans had, at most, a hundred or two more tanks than the French, and an uncomfortably high proportion of their units were equipped with the poorly-armed, almost obsolete PzKw I. The chief differences between the two sides were in organization and doctrine. The Germans concentrated their tanks in ten armoured divisions, organized in four armoured corps. About half the French tanks were allotted to independent battalions intended for infantry support. The rest were divided between four armoured divisions (one incomplete) and three light mechanized divisions. In other words, much of the French armour was so widely dispersed as to be almost useless in a campaign of rapid movement. Again, arms manufacturers cannot be blamed for an error of judgement for which the military authorities alone were responsible.

However, much of the information on which the foregoing comparisons are based did not become available to the Allies until the war was over. At the time they believed themselves to be heavily outnumbered in tanks and aircraft. One consequence was that the British were confronted almost from the first moment of the German offensive with urgent demands from the French for more fighters. These they met at the cost of reducing their fighter force at home to the equivalent of thirty-six or thirty-seven squadrons. At the same time, the British government and the Anglo-French Co-Ordinating Committee bombarded Washington with requests for American aircraft, to be found either from new production or from stocks held by the armed forces. On 15 May Churchill asked President Roosevelt for 'several hundred of the latest types of aircraft'. He also asked for 'the loan of forty or fifty of your older destroyers to bridge the gap between what we have now and the large new construction we put in hand at the beginning of the war.'[6]

Churchill's request for American aircraft was unrealistic for two reasons. In the first place, it was highly improbable that any aircraft the Americans might be able to send could arrive and be brought into action in time to affect the situation in France. (Indeed, of the aircraft already sent in response to orders placed in 1939 or earlier, many were still in their crates when the fighting stopped.) Secondly, the Americans had very few military or naval aircraft in stock, and the output of their aircraft industry was still extremely small. In the whole of the first six months of 1940 the British received only 104 aircraft from American factories. Their own factories produced 6,332.[7]

Within a few days of the despatch of Churchill's message,

the British Expeditionary Force and the French First Army were cut off from the main body of the French armies, now on the Somme–Aisne line. The British hoped to save up to a third of their troops by embarking them at Dunkirk or from neighbouring beaches, but all or most heavy equipment would have to be left behind. In the outcome, they managed to rescue very nearly the whole of the expeditionary force and a large number of French troops, but lost 880 field guns, 310 guns of larger calibre, about 500 anti-aircraft guns, 11,000 machine-guns, nearly 700 tanks and some 45,000 motor-cars and lorries. These losses—which left them with only enough modern equipment in the United Kingdom for about two divisions—would have to be made good from the output of British factories. The Americans had very little capacity for arms production, and none at all for the production of artillery of British pattern.

The British did, however, have in stock some 500 old 18-pounder field guns and 4·5-inch and 6-inch howitzers which could be pressed into service for home defence. They also had about 250 light and 230 cruiser and infantry tanks. The Americans were known to possess large stocks of rifles, machine-guns, mortars and 75-millimetre field guns left over from the First World War. Even before the fall of France, the Allies asked that some of these old weapons should be sent to them. The French intended to use the American field guns as anti-tank guns; the British needed small arms to replace losses and to equip the newly-formed Home Guard. The Germans were not expected to be ready for a full-scale invasion of the United Kingdom in the immediate future; but seaborne and airborne raids might, it was thought, do immense harm if local defence forces were not in a position to deal with them promptly.

The United States Chiefs of Staff responded to this request by producing a list of items they were willing to part with if the Department of State could find 'a legal way' of transferring them to Allied ownership. (To this list the President afterwards made additions.) Payment, the Allies said, should present no difficulty. A cable from London on 24 May made it clear that if necessary the British would commit the whole of their reserves of gold and dollars to the purchase of North American supplies.[8]

Eventually a method of transfer that satisfied the Anglo-Saxon legal mind was found. On 11 June 1940, the United States

Secretary of War, exercising powers conferred by an Act of Congress of 11 July 1919, sold the material to the United States Steel Export Company at a contract price of $37,619,556·60. Representatives of the United States Steel Export Company then took a taxi to the New York office of the British Purchasing Commission, where they resold the material at the same price to the British and French governments. A week later, France having capitulated, the parties agreed that the whole contract should be taken over by the British. The British also took over all outstanding contracts between the French government and American firms for the supply of airframes, aero-engines, machine-tools, explosives, raw materials and other items.

The material released included three quarters of a million of the rifles of Lee-Enfield pattern but ·30-inch calibre produced as a stop-gap by American small-arms manufacturers in 1917 and 1918; 895 75-millimetre field guns with carriages intended to be drawn by horses; and large numbers of machine-guns and Stokes mortars and some automatic rifles and revolvers. Items added after the contract was signed brought the total cost of the material to some forty-three million dollars.

Getting these weapons to their destinations proved no easy task. The American authorities went to a great deal of trouble to expedite their delivery from stores and arsenals in various parts of the United States to the army docks at Raritan, New Jersey, where the intention was that they should be taken aboard British (and originally also French) vessels and rushed across the Atlantic. Troops and civilians at Raritan worked far into the night week after week to prepare consignments for shipment, but found that facilities there were far from adequate.[9] Some consignments had to be diverted to other ports and left there to be picked up when shipping was available. Attempts were made to sort out consignments so that major components of dismantled weapons did not become separated and that slings, sights, bayonets and other detachable parts travelled with the appropriate weapons, but this was not always possible. Furthermore, it was perhaps less clear to the authorities in Britain than to the British and American authorities in the United States that, even where no parts or accessories were missing, the assembly of unfamiliar weapons whose metal components were packed in heavy grease was likely to prove a difficult and frustrating business for the recipients unless the services of American instructors or service manuals not always readily available could be provided.

A further complication was that there was no ·30-inch rifle ammunition in Britain and no capacity for its production. A certain amount of ·30-inch ammunition was included in the deal, but the British Purchasing Commission calculated that ten times as much would be needed. Millions of rounds from new production, which would have to be paid for in dollars, had therefore to be ordered, and clearance for these orders had to be obtained from American officials who were eager to help but worried about supplies for their own troops.

The Allied request for aircraft from stocks held by the United States Army and Navy was dealt with as a separate transaction. Again the Americans were willing to help, but the army and the navy had very few modern aircraft. They released 143 light bombers and dive-bombers. These were sold back to their manufacturers, who resold them to the Allied governments. Most of the dive-bombers, with some aircraft ordered earlier by the Belgian government, were taken aboard the French carrier *Béarn* at Halifax in the second week of June. The *Béarn* left Halifax in haste on the day of the French surrender, but was diverted by Admiral Darlan to Martinique, where she remained with the aircraft still aboard her until 1943. The light bombers were shipped from Halifax in a British carrier.

So these consignments, although generously offered and gratefully received, did not go by any means the whole way to provide the ready-to-use weapons and hundreds of modern aircraft the Allies thought they needed. Churchill, perhaps feeling that the material value of anything the Americans had to offer was less important than the benefits to be gained by aligning the United States behind the Allied cause, nevertheless asked for more. Roosevelt's reply to his request in May for the loan of forty or fifty old destroyers was to the effect that the time was not opportune for such a transfer. In the second week of June Churchill returned to the attack. Although he claimed that Britain's ocean trade might be strangled if thirty or forty American destroyers were not forthcoming, the question of the destroyers was linked in subsequent communications at the official level with requests for material needed for defence against invasion or seaborne raids. The Admiralty wished, in particular, to obtain the release of twenty-three motor torpedo boats of British design which were under construction for the United States Navy. On 18 June the Acting Secretary of the Navy aroused a storm of protest by announcing that twenty of these were to be released. Congress objected strongly to the transaction; the Attorney-General advised the government that it would be outside the existing law and that this would apply, too, to any transfer of destroyers.[10]

The matter was then allowed to drop for more than a month. On 22 July the British Ambassador in Washington, Lord Lothian, revived it by making, on his own initiative, a nation-wide broadcast in which he expressed the opinion that a hundred American destroyers and a few Catalina flying-boats might save Britain from invasion. When Churchill repeated his request for old destroyers at the end of the month and represented the matter as one of extreme urgency, President Roosevelt told Lothian that he could see no way of overcoming Congressional opposition to the transfer except by putting it forward as a transaction which would be beneficial to the United States and involve no risk of war.[11]

The sequel was the much-discussed 'destroyers-for-bases' deal. Some weeks before the outbreak of war the British had granted the United States government the right to use bases in Trinidad, Santa Lucia and Bermuda, and to acquire leaseholds and land stores there, in connection with the then still hypothetical neutrality patrol. A patrol by warships had been duly put into effect on the outbreak of war, but the right to use the bases had not been exercised because the flying-boats which were to have taken advantage of it were needed in the Pacific. The plan now concocted between the President, the Ambassador and Downing Street was that the British should receive the fifty destroyers and the twenty motor torpedo boats, in addition to fifty Liberators, five Catalinas, some dive-bombers, and a quarter of a million rifles for the Home Guard, and in return should grant the United States not the mere use of bases in three Atlantic islands, but ninety-nine-year leases of bases as far north as Canada and as far south as British Guiana.

However, at the last moment the British found that only the destroyers were to be released. This was not the result of an oversight, as was asserted after the war. The President had come to the conclusion that a straight destroyers-for-bases deal was the utmost Congress and the public would tolerate. He acknowledged a moral obligation to furnish the rest of the material, but said that the Attorney-General's ruling made it impossible for him to redeem his promise in full, at any rate for the time being.[12] The rifles were released some three weeks after the destroyers-for-bases deal was announced. Instead of the motor torpedo boats the British were offered additional aircraft, on the understanding that these would have to be paid for but that the money might be refunded later. Eventually they accepted some long-range bombers and the promise of fifty-five Catalinas to be delivered between November 1940 and the spring of 1941.[13] The motor torpedo boats were released in March. The destroyers did not arrive until after any danger of invasion in 1940 had receded, and a good deal had to be done to them before they were ready for use.[14]

In World War II the Tommy gun came into its own and made a fortune for the astute financer who had gained control of the enterprise

TWELVE

Arms and the 'Grand Alliance'

In the summer of 1940 British industry faced the task of replacing, as a matter of urgency, losses suffered in the Norwegian campaign and the campaign in France and the Low Countries. At the same time, the government's decision to safeguard the country's oil supplies and fulfil obligations to foreign governments by continuing to defend Malta, Egypt, Iraq, Palestine, Aden, the Sudan and Kenya meant that the British and Commonwealth forces in the Mediterranean and the Middle East had to be not merely supplied but strengthened.

As far as possible, deficiencies not arising from recent losses had also to be made good. The navy, for example, had many previously unforeseen requirements for small vessels, and these had somehow to be met at the cost of cancellations or indefinite postponements of long-term programmes. The Admiralty put its requirements under this head at nearly seven hundred escort vessels and fast minesweepers, five hundred magnetic minesweepers, not far short of two thousand trawlers, and some six hundred motor torpedo boats, motor launches and miscellaneous small craft.[1] Even when work on the big ships ordered since 1936 was suspended, Vickers-Armstrongs had naval orders on hand at Barrow and Newcastle alone to the value of some sixty million pounds.[2] Furthermore, a great many man-hours had to be expended on conversions and repairs to ships damaged by accident or bad weather. During the first four months of the war alone, well over a hundred ships had been damaged by such causes. This was about five times as many as had been damaged by enemy action.[3]

The Royal Air Force had lost since the spring not far short of a thousand aircraft, nearly half of them fighters. Representatives of the Air Staff and the newly-formed Ministry of Aircraft Production agreed before the end of May to prepare for the coming battle for air supremacy over Britain by directing aircraft manufacturers to concentrate their efforts, at any rate until October, on the production of Hurricanes, Spitfires, Blenheims, Wellingtons and Whitleys, even though this meant that the development and production of other aircraft had to be suspended.[4] Stimulated by a sense of urgency and by appeals for an all-out effort, the fighter factories stepped up deliveries from 256 aircraft in April and 325 in May to an average of 471 during the next four months.[5] In the autumn Fighter Command emerged victorious from the Battle of Britain with more aircraft than it had possessed at the beginning of the preliminary phase in July, while the effective strength of the German long-range bomber force declined by 200 aircraft between 10 August and the eve of the decisive phase on 7 September.[6] Between that date and the end of the month the Germans lost 433 aircraft destroyed or damaged beyond repair, the British 242.[7]

Replacement of the army's losses was a less spectacular and more long-drawn business. A few weeks before the Germans opened their assault on France and the Low Countries, the War Office had estimated its requirements for an expeditionary force of thirty-six divisions at some seven thousand tanks, well over twelve thousand new or converted field and anti-aircraft guns, and some thirteen thousand tank and anti-tank guns, to say nothing of enormous numbers of wheeled vehicles.[8] By the time the last British soldier returned from Dunkirk about seventeen hundred tanks, not far short of two thousand new field and anti-aircraft guns and some eighteen hundred new tank and anti-tank guns had been delivered, but meanwhile the number of divisions the government proposed to have ready for service in France by the end of 1941 had risen to the fifty-five tentatively proposed a good deal earlier.

At that stage the Ministry of Supply was confronted with the entirely new problem of equipping the numerically strong but poorly armed forces at home to resist invasion. The Home Guard (at first called the Local Defence Volunteers) could be armed largely with surplus weapons released by the United States Army, but still needed a good deal that could only be made in Britain. The full-time soldiers of the Home Forces command needed almost everything. Twelve divisions had returned almost empty-handed from France to join fifteen weak or incompletely trained divisions which were performing static tasks or working up for eventual despatch abroad.

In a sense, the problem was insoluble. There was no way in which the troops at home could be given enough mobility and hitting-power before the end of the campaigning season to enable them to drive out an invader who managed to come ashore in strength with artillery and armour. The safety of the United Kingdom depended in the summer of 1940 on the ability of the Royal Navy and the Royal Air Force to prevent this from happening. The authorities believed—correctly, as events proved—that the Germans, with only a few cruisers and destroyers and at most two battle-cruisers and two old battleships at their disposal, would not attempt a full-scale landing without first fighting a battle for air supremacy which the Royal Air Force would win.

The enormity of the task that faced the Ministry of Supply and the arms industry in June is starkly revealed by a glance at the state of affairs in Home Forces about the time when the first troops were coming back from Dunkirk. The part of England most likely to be invaded and most exposed to seaborne raids was the stretch of coast between the Wash and Selsey Bill. This was guarded on 31 May by six infantry divisions of Eastern Command. Six infantry divisions ought to have had 432 25-pounder field guns and 288 anti-tank guns. Eastern Command's divisions had 39 and 14 respectively. These were supplemented by 124 superseded 18-pounders and 4·5-inch howitzers.[9]

Neither the Ministry of Supply nor the arsenals and arms factories that served it can be justly blamed for these shortcomings. An appallingly late start had been made with the rearmament of the land forces, but that was not their fault. Substantial numbers of new 25-pounders and converted 18-pounders had been delivered. So had substantial numbers of 2-pounder tank and anti-tank guns. Procurers and suppliers of arms could not be expected to foresee that the expeditionary force would lose most of them.

Nor can the framers of production programmes for guns and ammunition be accused of doing a bad job. They had gone to great pains to ensure that the muddles and mistakes of World War I would not be repeated. Recognizing that specialist firms ought not to have been burdened in 1914 and 1915 with orders for shells and other stores which could be manufactured by general engineering firms, they planned for the new and improved war a rational division of tasks between government factories, commercial arms manufacturers, commercial firms outside the arms industry, and 'agency' or shadow factories owned by the state but managed by private enterprise. In principle, the barrels and mechanisms of the larger guns would be made in government factories and by commercial arms manufacturers (which in effect meant Vickers-Armstrongs); shell-cases, carriages and mountings by outside firms. The outside firms would also make important contributions to the conversion of 18-pounders. Shell-filling, which bore no resemblance to any normal commercial process, would be undertaken by government factories, including the one national factory of the 1914–1918 era still in existence and a number of new factories to be set up before and immediately after the outbreak of war. The manufacture of explosives, so far as United Kingdom production was concerned, would be a task for government factories, Imperial Chemical Industries and agency factories. The first orders for 2-pounder tank and anti-tank guns would go to the Royal Arsenal at Woolwich, Vickers-Armstrongs and a new government factory, but later non-specialist firms would be expected to take a hand in their production. Small arms were to be made initially at Enfield and by such experienced firms as Vickers-Armstrongs and BSA. Manufacture of the Bofors light anti-aircraft gun in the United Kingdom would be entrusted to Nuffield Mechanizations and Aero Limited and a new government factory specially equipped for the production of such weapons.

On the whole this system worked very well. When the 20-millimetre Hispano-Suiza was adopted in 1938 a new firm, the British Manufacture and Research Company, was invited to join government factories and other commercial firms in producing it under licence, and as time went on the system was modified in the direction of participation by commercial firms outside the arms industry in the manufacture of a wider range of weapons. The fact remains that the system was designed to provide arms and ammunition for a small expeditionary force on the outbreak of war and a larger one some two years later, not to meet an unforeseen emergency some nine months after hostilities began.

Even so, some progress was made. Deliveries of new 25-pounders rose from an average of 42 a month in May and June to 70 a month in the next three months. Thereafter it increased markedly, reaching an average of well over 500 a month in the last three months of 1941. In May 1940, 126 2-pounder tank and anti-tank guns were produced; in June 169; in July, August and September an average of 166 a month. This modest output gave Home Forces 498 anti-tank guns at the end of August, as compared with 176 in June. In 1941 production started at 281 in January and soared to nearly 1,400 a month in November and December.

Deliveries of wheeled and tracked vehicles were not so satisfactory. From May 1940 to December 1941, some 5,400 to 9,600 lorries, trucks, ambulances and large motor-cars left the factories each month, but no steady upward trend was visible. As for tanks, a War Cabinet committee made in April 1940 the stultifying pronouncement that the tank programme 'must not be interfered with either by the incorporation of improvements . . . or by the production of newer models'. Neither the Ministry of Supply nor the War Office took this prohibition very seriously, but after the withdrawal from Dunkirk immediate priority had perforce to be given to the production of existing types. Tank manufacturers had, however, to compete with aircraft manufacturers for machine-tools and gauges, and to some extent for partly-processed materials and skilled labour. Output increased very slowly in 1940 from 113 medium and heavy tanks in May and 115 in June to an average of 136 a month in the last half of the year. It varied between 200 and some 240 a month in the first half of 1941, but then climbed to more than 600 a month as the sequel to a ruling that tanks should have the same priority as was given in the previous year to aircraft.

*

Even at the height of the invasion scare, long-term plans were seldom far from the thoughts of the Churchill government and its military advisers.

Italy's entry into the war had, however, the serious disadvantage for the British of forcing them to divert their shipping bound to or from the Middle East to the long route round the Cape. Occasional fast convoys could be rushed through the

Mediterranean, but only at great risk. The Germans could, and did, send aircraft to Sicily to join the Italians in asserting control of the Sicilian Narrows, and later they sent troops and aircraft to North Africa. Delays arising from the use of the Cape route reduced the effective carrying capacity of the British merchant fleet at a time when the Germans were expanding their U-boat force for the purpose of reducing it still further.

Even so, the British did not intend to stand permanently on the defensive. Their broad aims after the Battle of Britain was won were to win the coming Battle of the Atlantic; reopen the Mediterranean to their shipping by defeating the Axis powers in Libya; and build up their strength at home for a return to the mainland of Europe about the end of 1942 or early in 1943. In the meantime bombing and economic blockade would, they hoped, wear the Germans down to a point at which they would succumb to the combined effects of uprisings in the occupied countries, political dissent in Germany itself, and the arrival of a British expeditionary force. The 55-division project was retained as a basis for production and procurement programmes, not because the government intended to put 55 divisions ashore in Europe but because the needs of home defence and the Middle East had to be taken into account.

In the summer of 1940, much thought was given to the possibility of drawing on American productive capacity to sustain the British effort. The British were eager buyers of American machine-tools, raw materials and food, but as yet had made comparatively small purchases of American munitions. Of all munitions supplied to British and Commonwealth forces during the first sixteen months of the war, less than six per cent came from the United States, nearly ninety-one per cent from the United Kingdom, between three and four per cent from Canada, Australia, New Zealand and India.[10]

In May 1940, the United States government contemplated duplicating British and American factories in the United States as an insurance against their destruction by bombing. But American attitudes changed markedly at the time of the withdrawal from Dunkirk. Three days after the withdrawal was completed, the Joint Army and Navy Board adopted as a basis for strategic and logistic planning a staff study which assumed that Britain and France would be defeated and that the United States and Canada might face the combined strength of Germany, Italy and Japan. President and Congress went on to accept an ambitious programme of industrial expansion and selective military service intended to raise and equip by the end of 1941 an army of two million men. In the worst case, the United States would withdraw her forces in the Pacific to the triangle Panama–Hawaii–Alaska, station most of the fleet in the Caribbean, and try to defend the Western hemisphere only as far south as Bahia.[11]

However, the worst case assumed that the Germans would capture major elements of the British and French fleets. The Americans were heartened by the news that no French capital ships had fallen into German hands, by an assurance that the British would rather send their fleet to Canada than allow the Germans to take possession of it, by British successes in the Battle of Britain. Moreover, their experts were powerfully impressed by the unconditional disclosure by a British mission in August and September of scientific and technical information of incalculable military value. The British had already handed the President the complete designs and specifications of the Rolls-Royce Merlin engine, leaving it to him to decide what return he wished to make for the right to manufacture it and for help in getting its production by the Packard Motor Car Company under way. The information now disclosed and the radar and sonar equipment shown opened new vistas to the American scientists to whom their secrets were imparted. This equipment included the cavity magnetron valve, afterwards described by an American official historian as 'the most valuable cargo ever brought to our shores'.[12]

In the meantime a great many man-hours were devoted in London and Washington to discussion of the extent to which the British should seek, and should be allowed, to draw upon an expanding American output of war material. The British still tended to think of American productive capacity as a reserve to which they might turn in an emergency rather than an immediate source of the tanks, guns and other equipment they would need for their return to the Continent. At the same time, they saw the force of the argument that failure to place orders for munitions in good time might mean that such orders could never be placed, and worse still might prejudice their chances of continuing to receive the machine-tools and raw materials they could not afford to be without. Similarly, in the United States there were two schools of thought with respect to the advantages or disadvantages of allowing the British a share of the national output. The President argued that arming the enemies of totalitarianism was good policy and sound strategy. The Chiefs of Staff accepted with misgivings the President's additions to their list of surplus material, considered that the British ought not to be allowed to place orders which might hamper American rearmament, and at first were doubtful about allowing them to receive all the aircraft due to them under the contracts taken over from the French. In July, however, they came round to the view that the best solution would be to expand aircraft production so that both American needs and the foreign contracts could be met. In practice production did not come up to expectations, and output was shared in accordance with a series of *ad hoc* decisions by a committee of American and British officers and officials appointed in the first instance for the purpose of making agreed adjustments to production schedules and trying to arrive at common standards.

The setting up of this committee was a step towards a kind of partnership helpful to both parties. The British gained access to American supplies, the Americans gained the benefit of British experience of combat conditions in Europe and Africa. Such arrangements worked fairly well where aircraft were concerned, better still with tanks. At the end of July a British Tank Mission reported favourably on the American M-3 medium tank (afterwards called the General Lee), adding that the War Department was already committed to its manufacture. The British government, confronted with the choice between ordering the M-3 and renouncing any immediate prospect of obtaining tanks from the United States, placed an initial order for 1,500. Once this was accepted, the British had little difficulty in persuading the Americans to make modifications which their experience had shown to be essential. Thereafter the British and the Americans co-operated closely in the development of armoured fighting vehicles for both armies.

Artillery and small arms were another matter. In August 1940, the War Office put its requirements under the 55-division scheme at more than 10,000 tanks, about 6,000 field guns, some 1,400 new or converted medium guns, more than 15,000 anti-aircraft guns and some 20,000 tank and anti-tank guns, in addition to large numbers of wheeled vehicles. Requirements formulated later for equipment to be available by the end of 1942 were still larger. Where the equipment was to come from was not altogether clear. The British Army had no objection to mixing American with British tanks, and here American production would be invaluable. But would the United States government, whose own army used 105-millimetre field guns, 90-millimetre and 37-millimetre anti-aircraft guns and 37-millimetre and 75-millimetre tank and anti-tank guns, allow American arms factories to manufacture for the British 25-pounder field guns with a calibre of 3·45 inches; 3·7-inch, 4·5-inch and 40-millimetre anti-aircraft guns; and 40-millimetre 2-pounder or 57-millimetre 6-pounder tank and anti-tank guns?

In September 1940, Sir Walter Layton, Director-General of Programmes in the Ministry of Supply, crossed the Atlantic to put this question to the authorities in Washington. He took with him a list of weapons needed to plug gaps in the British production programme or insure against a serious loss of output as a result of enemy action. These included 1,800 field guns; 2,250 tank-guns for British-built and 3,000 for American-built tanks; at least 1,600 heavy and 1,800 light anti-aircraft guns; 2,000 2-pounder and 1,000 6-pounder anti-tank guns; and a million ·303-inch rifles. Rifles of the American ·30-inch calibre would do for the Home Guard, but not for permanently embodied troops.

In one respect the answer was unequivocal: the United States government would not finance the installation of plant for the manufacture of weapons not used by American forces. This ruled out 25-pounder field guns, unless the British were willing to meet the initial cost of their development in the United States and to wait until the middle of 1941 for the necessary machine-tools.[13] It also ruled out any production of 3·7-inch anti-aircraft guns within the next two years. As for Layton's other requests, the authorities made it clear that they would be inclined to look more favourably on these if the British would help their production programme by accepting an offer of American-type equipment for a complete force of ten divisions to be raised, trained and maintained in the field independently of the main body of the British Army.

From what source the ten divisions were to draw their manpower was not clear. Without, apparently, bothering about that aspect of the matter, the British government authorized Layton to accept the offer. The American authorities then sanctioned orders for 2-pounder and 6-pounder anti-tank guns and for 4·5-inch medium guns (not urgently required by the British), since these were weapons the United States Army was thinking of employing. They also allowed the British to take over and lease to the Remington company for the manufacture of ·303-inch rifles a disused national arsenal at Rock Island. Finally, they offered to facilitate Canadian production of 25-pounder field guns by allowing American firms to act as sub-contractors.

The ten-division project was short-lived. By the time Layton's two-and-a-half-months' stay in the United States ended, the American Chiefs of Staff were no longer thinking in terms of the last-ditch defensive strategy adopted in June. Apart from the political aspect, there were at least two good reasons for this. One was the slow pace of American rearmament. By October it was clear to the Chiefs of Staff that there was no prospect of their being able to equip an army of even 1,400,000 men—let alone two million—before the spring of 1942. The other was that Britain was still undefeated, her fleet still intact. The security of the Western hemisphere, the Chiefs of Staff now thought, depended more on their remaining so than on any other single factor. General George C. Marshall, Chief of Staff of the United States Army, felt by the end of October that it was imperative to 'resist proposals that do not have for their immediate goal the survival of the British Empire and the defeat of Germany'.[14] Admiral Harold R. Stark, his naval counterpart, went even further. Reviewing courses of action open to the United States in a memorandum written on 4 November and afterwards revised, he came to the conclusion that the armed forces must direct their efforts towards 'an eventual strong offensive in the Atlantic as an ally of the British, and a defensive in the Pacific'.[15] 'The issues in the Orient,' the Joint Planning Committee of the Army and Navy Board wrote on 21 December, 'will largely be decided in Europe.'[16]

The re-election of Franklin D. Roosevelt as President went a long way to solve the problem of securing popular support for the policy of aid for Britain with which he was associated. There remained the difficulty that aid for Britain meant selling the

British food, raw materials, machine-tools and munitions they might soon be unable to afford. By the latter part of 1940 it was clear that their dollar resources would not last very much longer, even at the current rate of expenditure. If all the orders the British contemplated placing and the Americans accepting at the end of the year were added to orders already in hand, the British would not be able to pay for more than about half the goods supplied. This would not be a mere embarrassment to the British and their suppliers. It would defeat the strategy the United States Chiefs of Staff had just decided to adopt.

The President turned this problem over in his mind while cruising in the Caribbean in December. He returned with the idea of lend-lease. The sequel was the passing on 11 March 1941 of an act which empowered the President to authorize the Secretary of War, the Secretary of the Navy, or the head of any other government department or agency to manufacture in arsenals, factories and shipyards under their jurisdiction any defence article for the government of any country whose defence the President deemed vital to the defence of the United States.

In one of the informal broadcasts to the nation which Roosevelt was pleased to call his 'fireside chats', he likened lend-lease to a hose one might lend to a neighbour to enable him to put out a fire before it spread to one's own house. As many writers have pointed out, the comparison was not altogether apt, because the munitions sent to the British were not going to be returned when the conflagration was over. Lend-lease might be more aptly compared with a club given by a man to a friend for the purpose of knocking a mutual enemy on the head. To discard metaphor, it was a device which enabled the United States to wage an undeclared war by helping the British to fight the Germans and the Italians until her own armed forces were ready to enter the fray.

The help given to the British during the period of American non-belligerency was, however, very small. Lend-lease did not relieve them of the necessity of paying for goods ordered before the act came into force, and in fact they paid in gold or dollars for nearly all material delivered up to the time when the United States entered the war. In the whole of 1941 about eleven-and-a-half per cent of all munitions furnished to British and Commonwealth forces came from the United States, and only about a quarter of this material was supplied under lend-lease arrangements.[17]

Early in 1941, after the decision to introduce lend-lease had been taken but before the act was passed, Anglo-American staff talks opened in profound secrecy in Washington. The British and the Americans agreed that, if Japan entered the war, they should stand on the defensive in the Far East and the Pacific until Germany was defeated, but differed as to the means by which the defeat of Germany was to be brought about. General Marshall had little faith in the British policy of wearing the Germans down by bombing and blockade. He believed that only direct intervention by powerful American forces would suffice. The crucial question, however, was whether the means of carrying powerful American forces safely across the Atlantic could be found. Many more ships than were available in the early part of 1941 would be needed, and in any case the Battle of the Atlantic must first be won.

Neither finding the ships nor winning the Battle of the Atlantic promised to be easy. By March British, Allied and neutral merchant shipping was being sunk by German submarines or surface raiders or lost from other causes at the rate of well over half a million tons a month. The British shipbuilding industry, hard hit by a long slump, was in no position to replace such losses, and American shipbuilding capacity was still quite small. In 1939 American shipyards had built only twenty-eight ocean-going ships, in 1940 only fifty-three. Few of the sixty ships ordered by the British were likely to be ready before the end of 1941, and in fact only five of them were completed within that time. Altogether about a hundred merchant vessels of all classes came from American shipyards in the course of the year.

Steps were taken on both sides of the Atlantic not only to increase output but also to reduce losses. The American authorities drew up production schedules which envisaged the completion of more than 1,200 ships by various dates before the end of 1943. In April the President gave orders that American warships should patrol shipping routes in the Western half of the Atlantic. In May reinforcements, including three battleships, were brought from the Pacific. The British made a successful attempt to cut down sinkings by routing convoys further to the north, increasing the range and strength of surface and air escorts and making more and more use of improved radar devices. New bases for long-range aircraft in Northern Ireland, the Hebrides and Iceland were brought into service. By the end of May British and Canadian escort vessels were covering the full width of the Atlantic. By midsummer what seemed to be the battle, but in fact was only the first phase of it, was won. Losses were very much lower in the second than in the first half of the year. Indeed, they were lower, in terms of the monthly average, than those suffered in the first sixteen months of the war.

In the meantime the Germans introduced a new factor by invading Russia. The British, alive to the importance of not allowing the Russians to be defeated, promptly sent them 450 aircraft, 22,000 tons of rubber, three million pairs of boots and substantial quantities of aluminium, tin, lead, jute and wool.[18] The Americans decided in the light of a report brought back from Moscow by the President's personal emissary, Harry Hopkins, that they, too, would help the Russians to stay in the war.

The Russians then drew up a list of their requirements for the period from October 1941 to the following June. These included 4,500 tanks and 3,600 aircraft. The British promised 2,250 and 1,800 respectively; the other half, they assumed, would be forthcoming from the United States. In the middle of September the Americans announced, however, that they could find only 1,524 tanks, not 2,250, and that more than half of them would be tanks hitherto intended for the British.[19] The British thereupon asked them to provide the remaining 726 at the cost of still further reducing their allocation. Similarly, the Americans at first offered only 1,200 aircraft instead of 1,800.[20] Eventually they agreed to find the other 600 by drawing on the allocation to their own armed forces, but said that the British allocation would have to be adjusted later.

As a result of these transactions, the build-up of British strength in the Middle East was seriously delayed. On the other hand, the British and the Americans had the satisfaction of being able to tell Stalin that his needs would be fully met so far as tanks and aircraft were concerned, and an agreement to that effect (the 'first protocol') was signed in Moscow on 2 October 1941. In addition to the 2,250 tanks and 1,800 aircraft on which the British insisted, the Americans promised 152 anti-aircraft and

756 anti-tank guns and large quantities of other military stores.[21]

But their satisfaction was short-lived. American production in the first few of the nine months covered by the agreement was far too small to provide the weapons promised to the Russians and meet other commitments. The result was that, although the British were able from the outset to redeem their promises, the Americans were not. Eventually 2,249 of the promised 2,250 tanks, four of the promised 152 anti-aircraft guns and sixty-three of the promised 756 anti-tank guns were shipped to the Soviet Union, and 1,727 of the promised 1,800 aircraft were made available, although only 1,285 were shipped.[22] But the situation some two months after the signature of the first protocol seemed well-nigh desperate.[23] Only about a quarter of the material promised to the Russians within the first two months had been shipped, carrying capacity for cargoes consigned to the Soviet Union was hard to find, and Russian officers sent to the United States to supervise shipments were rejecting a high proportion of the aircraft offered to them on the ground that they were incomplete, unsuitably packed or otherwise defective. To make even meagre shipments to the Russians possible, deliveries to the British had been slashed and allocations to United States forces at home and abroad cut to the bone. Although the defence of China had been declared essential to the security of the United States, very little war material had been sent to the Chinese in recent months. Moreover, there was reason to suspect that only about half the material the Americans were able to send to Kunming by the Burma Road was reaching its destination. The Dutch authorities in the Netherlands East Indies, although able and willing to pay in dollars for American small arms and anti-aircraft guns, were short of both, and British and American garrisons in the Far East and the South-West Pacific were far from adequately armed. It was hoped, however, that by the spring of 1942 the British would have a strong fleet at Singapore and the Americans about a hundred B-17 heavy bombers in Luzon.

*

About the time of the Anglo-American Atlantic Conference in August 1941, a number of American officers had the impression that the British, and to some extent the President, were trying to push the United States towards war. In view of the slow pace of

American rearmament, the shortage of shipping and the precariousness of the British hold on Egypt such pressures, it was felt, should be resisted. According to current calculations, United States forces would not be strong enough before 1943 to make a useful contribution to a campaign in Europe or Africa.

There was also an impression in the United States that the Japanese were preparing to take the offensive against the Western Powers in the near future. That this belief was widely held is understandable. In 1940 a pro-German Foreign Minister, Yosuke Matsuoka, had persuaded a dubious Privy Council to sanction a tripartite pact between Japan, Germany and Italy. On 2 July 1941, the Japanese had decided to move troops into southern Indo-China for the purpose of seizing bases for a possible invasion of the Netherlands East Indies. These movements had begun about the middle of July, and the United States government had responded to them by putting an embargo on trade with Japan and asking the British and the Dutch to take similar steps.

By August, however, the Japanese had rid themselves of Matsuoka and were making strenuous attempts to improve their relations with the West. Negotiations in which the American Secretary of State, Cordell Hull, played a leading part culminated on 7 November in proposals by the Japanese for a comprehensive settlement of Far Eastern questions. When Cordell Hull replied that public opinion in the United States would not tolerate a comprehensive settlement with Japan while the tripartite pact remained in force, the Japanese suggested an interim settlement on terms which would include an undertaking on their part to make no further advance in Indo-China. On 22 November Hull showed their proposals to representatives of the United Kingdom, Australian, Chinese and Netherlands governments and invited comments on counter-proposals drafted in the Department of State. On 26 November, however, he abruptly gave up the whole idea of an interim settlement. With the President's concurrence, he then confronted the Japanese with an offer to treat only if they withdrew their forces from the whole of China and Indo-China and relinquished their extra-territorial rights. As the result of an oversight, he omitted to add that these demands were not meant to apply to Manchuria.[24]

Since these terms were not acceptable to the Japanese, the result was that Britain and the United States found themselves committed on 7 December to a war for which they were far from ready. Eight of the nine American battleships at Pearl Harbor were sunk or crippled, and Malaya was invaded, while the Japanese Embassy in Washington was still struggling to 'put in nicely drafted form' a long message breaking off the negotiations.

Just over a fortnight later Winston Churchill, accompanied by the British Chiefs of Staff and the Minister of Supply, arrived in Washington for a series of meetings with President Roosevelt and the United States Joint Chiefs of Staff. One of the first decisions reached was that Britain and the United States should stick to their intention of standing on the defensive in the Far East and the Pacific until Germany was defeated. Attempts to check the Japanese advance and secure bases for an eventual offensive were destined, however, to make heavy demands on their resources. Between December 1941 and June 1943—a period which covers the Anglo-American landings in North-West Africa and a big build-up of American forces in Britain—about a third of all troops which left the United States in ships controlled by the United States Army went to the Central, South or South-West Pacific.[25]

Another important consequence of the meetings was that Anglo-American bodies were set up to control strategy, procurement and supply. To avoid confusion with such purely American bodies as the Joint Army and Navy Board, these were called not joint but 'combined' committees or boards. The framing of strategy, under the broad direction of the President and the Prime Minister, was entrusted to a Combined Chiefs of Staff Committee. In theory this consisted of the American Joint Chiefs of Staff and the British Chiefs of Staff in joint session. In practice the British Chiefs of Staff could sit with their American colleagues only at summit conferences. They were therefore represented in Washington by delegates drawn from the British Joint Staff Mission. The Combined Chiefs were served by a combined secretariat and a small combined planning staff, but most planning continued to be done on a national basis. Papers were prepared by the staffs in London and Washington and submitted to the Combined Chiefs for examination and decision. A directive given to the Combined Chiefs of Staff instructed them to effect 'the collaboration incident to their responsibilities' by making use of 'appropriate combined or other bodies in Washington, London, and elsewhere'.[26] A Combined Raw Materials Board, a Combined Munitions Assignment Board and a Combined Shipping Adjustment Board were set up on 14 January 1942, a Combined Production and Resources Board and a Combined Food Board on 9 June.

The Allies thus provided themselves with what promised to be an admirable means of ensuring that their vast resources were used for the common good. As things turned out, the machine worked far from smoothly. There were many reasons for this. One was that the President and the Joint Chiefs of Staff on the one hand, and the Prime Minister and the British Chiefs of Staff on the other, could not always agree about the strategy to be pursued. They differed on more occasions than one, for example, about the relative importance of the Mediterranean and North-West European theatres and of the Far Eastern and Pacific theatres. Another was that the Combined Chiefs of Staff, as a body, were able to exercise only a loose control over American forces in the active theatres of war. Even when they did bring themselves to issue firm directives, these were sometimes misinterpreted, successfully challenged, or flatly disregarded by Commanders-in-Chief or Commanding Generals. A third reason was that the Combined Chiefs made little use of their theoretical power to control the production and distribution of material resources by bringing pressure to bear on supply and procurement agencies through the combined board. American thinking about these matters tended to draw a sharp distinction between military and civilian functions. With the exception of the Combined Munitions Assignment Board, none of the combined boards had much direct contact with the Combined Chiefs or received from them the strategic guidance needed to make the system work as effectively as it should have done. Inadequate foreknowledge on the part of the supply departments of strategic needs was, for example, the most important single cause of the persistent shortages of landing craft which plagued the Allies in 1943 and 1944.

The Combined Raw Materials Board was formed by the bringing together of W. L. Blatt of the United States Office of Production Management and Sir Clive Baillieu of the British Raw Materials Mission in Washington. Its first task was to deal with the situation created by the imminent loss of Malaya, the

Indonesian archipelago and adjacent territories. These produced about ninety per cent of the world's rubber and sixty per cent of the world's tin. Elsewhere in the Far East and the South-West Pacific the Allies were soon to lose important sources of tungsten, chromite, antimony and hardwood and the sole source of manila hemp. The board tackled, with conspicuous success, the problems that stemmed from these losses by seeking out stocks, developing alternative sources of supply, finding substitutes for scarce materials and fostering the manufacture of synthetic products. The British agreed that the United States should undertake the production of synthetic rubber for both countries.

Steel and aluminium—key materials for manufacturers of arms and aircraft—were left to be dealt with by the existing agencies. Deliveries of steel depended on production, allocation and shipping capacity rather than the supply of raw materials, and these were felt not to be matters for the Combined Raw Materials Board. Supplies of aluminium from bauxite extracted in the Western hemisphere were stepped up by increased British investment in the Canadian aluminium industry and additions to productive capacity in the United States. The United Kingdom was destined to import substantial quantities of fabricated aluminium from the United States in the latter part of the war, but these imports were offset by the export of aluminium ingots from Canada.

The Combined Munitions Assignments Board was intended to do for munitions in general what the Joint Aircraft Committee had hitherto done for aircraft. Unlike most of the combined boards it consisted not of one American and one British member but of Harry Hopkins as civilian Chairman and three officers from the armed forces of each country. It also differed from the other combined boards in securing some degree of co-operation from the Combined Chiefs of Staff. Its mandate was to allocate equipment in accordance with the principle that 'the entire munitions resources of Great Britain and the United States' should be deemed to be a common pool and that munitions should be allocated to the forces that could make the best use of them, irrespective of their country of origin or the national status of the agency that procured them. By virtue of this mandate the board was responsible for allocations not only to Britain and the United States but also to other Allied countries—except the

Soviet Union, whose allocations were fixed by protocol—and to non-belligerent countries whose interests were regarded as in some measure bound up with those of the Allies. Since this was too big a task for a single board in Washington to tackle, authority was given to a London Assignments Board to make allocations subject, in theory, to revision by the main board. The London board was headed by Oliver Lyttelton, who succeeded Lord Beaverbrook as Minister of Production soon after the post was created in February 1942. The Washington board made the main allocations and also made sub-allocations to the Latin-American countries and to China. The London board made sub-allocations to the British dominions and colonies and to Egypt and Turkey.

The Combined Shipping Adjustment Board consisted of Sir Arthur Salter, Chairman of the British North American Supply Committee, and Admiral Emory S. Land of the United States Maritime Commission. Their instructions were to 'adjust and concert in one harmonious policy' the activities of the British Ministry of War Transport and 'the Shipping authorities of the United States government'. They had no executive power, and no single authority in the United States exercised over American shipping a degree of control comparable with that exercised over British shipping by the Ministry of War Transport. The Maritime Commission was not a distributive but a procurement agency, concerned with the production of merchant vessels. From the moment when the United States became an active belligerent, the Allies faced problems in regard to shipping which the organization at their disposal did not allow them to solve by practising the rigid economy forced on the British when they fought alone. Eventually a vast increase in American shipbuilding capacity helped to overcome their difficulties, but they went through anxious times before winning the second round of the Battle of the Atlantic in the late spring of 1943 and reopening the Mediterranean to Allied convoys later in the year.

The Combined Production and Resources Board was first mooted early in 1942. In the United States, a War Production Board assumed ultimate responsibility in January for all American production for United States forces. In Britain the functions of the Ministry of Production set up in February were to make broad plans for future production in the United Kingdom and to co-ordinate transactions with the United States government and the combined boards. By the spring it was common ground between the British and the Americans that both must modify their production programmes in the light of an agreed strategic plan. In April Donald Nelson, Chairman of the War Production Board, wrote to the British Supply Council to say that production in the United States was running far behind schedule and that more realistic targets must be set.[27] He suggested that British and American representatives should join in reviewing the programmes of both countries. The outcome was the setting up of the Combined Production and Resources Board, consisting of Nelson and his opposite number Oliver Lyttelton and of an American and a British deputy. A directive issued by Roosevelt and Churchill made the board responsible for combining the production programmes of Britain, the United States and Canada into a single integrated programme 'adjusted to the strategic requirements of the war'. The Combined Chiefs of Staff and the Combined Munitions Assignment Board were to be responsible for keeping the new board 'currently informed concerning military requirements'.[28]

At its first meeting, held in Washington on 17 June, the Combined Production and Resources Board formally invited the Combined Chiefs of Staff to instruct the service authorities to prepare two statements: one showing what munitions the Combined Chiefs of Staff required by the end of 1942, the other the scale and character of the British and American forces to be equipped and deployed in all theatres in the spring of 1944. At a meeting of the Combined Chiefs of Staff on the following day, Lyttelton explained that a Combined Order of Battle of the forces that would be available for the campaigns of 1944, translatable into terms of munitions, was needed to enable the Combined Production and Resources Board to frame its production programme for 1943 in accordance with strategic requirements, as it had been ordered by the President and the Prime Minister to do. The Combined Chiefs accepted the board's invitation and told their planning staffs to get to work.[29]

Field-Marshal Sir John Dill, Churchill's personal representative in Washington, predicted in the light of his knowledge of the American administrative machine that the United States Joint Chiefs of Staff would be unable to bring themselves to work with a civilian agency and that therefore the information needed for the Combined Order of Battle would not be forthcoming. Lyttelton, a member of the British War Cabinet and a one-time professional soldier with a distinguished record of service, found this almost incredible. He left Washington and returned to London believing that all was well.

But Dill was right. There was so much delay in getting out the figures for 1942 that in August the board decided to concentrate on the information needed to determine the production programme for 1943. On 19 August President Roosevelt undertook to see that American requirements were stated; the Combined Chiefs of Staff agreed on 28 August to make every effort to provide the data. They had no difficulty in obtaining the British contribution to the Combined Order of Battle, but the American contribution was still to seek. Finally, on 8 September the board learned that the United States Joint Chiefs of Staff had declined to produce it, on the ground that the preparation of such a document at so early a stage was not feasible. They proposed that their requirements should be calculated on the basis of a forecast of the forces that could be transported and maintained overseas and of those they would need to retain in the Western hemisphere for defensive purposes, for training or as strategic reserves.[30]

The attempt to frame an integrated production programme related to strategic needs in active theatres of war had therefore to be abandoned. The Combined Production and Resources Board had to relinquish its primary function and turn to its secondary task of so adjusting British and American output as to avoid wasteful duplication and of achieving a satisfactory balance between production for military purposes and production to meet civilian needs.

In the meantime some rather alarming discoveries were made. The British had agreed earlier in the year that, since the munitions output of the United States was about to overtake Britain's and would soon surpass it, the United States should henceforth provide both countries with all their forty-ton tank transporters and ten-ton lorries, very nearly all their transport aircraft and self-propelled artillery, and a high proportion of their merchant vessels, auxiliary carriers and other escort vessels. These arrangements were made at a time when the looseness of the control the United States government was able to exercise over American procurement agencies was not fully apparent to the British. They still believed that the system of combined boards would work and that it would enable the Combined Chiefs of Staff to ensure that productive capacity was used to the best advantage for the common good. But towards the end of 1942 some disconcerting consequences of the inability of the American War Production Board and the Combined Production and Resources Board to do all that was hoped of them came to light. The United States Department of the Navy had placed large orders for landing-craft with seventy-nine different firms, most of them with no previous experience of the work they were undertaking and many a long way from the sea.[31] Only a last-minute scramble and the British contribution enabled the Allies to find enough landing-craft for their expedition to North-West Africa in November, and even then the landings had to be postponed and their scope reduced. It was also found that orders had been placed for more than a hundred times as much ball ammunition for use in 1943 as had been used throughout the campaigns in the Middle East up to the autumn of 1942 and for enough tanks to give 200 armoured divisions 225 tanks each and provide 100 per cent reserves.[32]

Such extravagances might seem harmless in view of the vast war potential of the Allies. But they would not be harmless if they led to a situation in which the British found themselves short of tank transporters because the Americans had built more tanks than could be used.

Moreover, as a result of the growing military strength of the United States and the extension of lend-lease to countries other than Britain, even well-informed Americans tended by the latter part of 1942 to forget that lend-lease was an arrangement made for the mutual benefit of the British and United States governments, and to regard it as a kind of dole to be granted or withheld at will. Protocols renewed annually from the summer of 1942 assured the Russians that requirements stated to and accepted by the Allies would be met and that their right to receive military stores and information of great military and industrial value would not be questioned. The British had no such guarantee. American arms programmes for 1943 added up to a total which exceeded by some 24 per cent the War Production Board's estimate of the maximum attainable production. Inevitable cuts seemed to the British sure to be made largely at the expense of lend-lease assignments and more than likely to affect their allocations of aircraft, merchant shipping and escort vessels. Even before learning in October that their allocation of aircraft was in fact to be reduced, they decided in principle to send a mission to Washington for the purpose of reminding the United States government of the importance of examining British and American production programmes as a whole.

The mission, headed by Oliver Lyttelton, reached Washington on 4 November and stayed until the end of the month. Its arrival coincided with the climax of the Second Battle of Alamein and preceded by a few days the Anglo-American landings in North-West Africa. Lyttelton found that the Americans were now thinking in terms of a twenty-five per cent cut in allocations to their own armed forces and that this might entail a corresponding cut in Britain's lend-lease assignment. Although this would be a serious blow, he was concerned not so much to avoid a cut as to obtain an assurance that anything promised after the cut was made would in fact be forthcoming. About thirty per cent of the adult population of the United Kingdom was already serving in the armed forces or employed in the civil defence organizations or the munitions industry. The government would soon have to make up its mind how to divide its last reserves of manpower between the fighting services and the factories. It could not make a rational decision without knowing how much of the common pool of munitions Britain was to receive.

Lyttelton returned to London at the end of the month with a letter from President Roosevelt to the Prime Minister which contained a number of assurances. Some of them were so hedged about with qualifications that their worthlessness was self-evident. Experience during the rest of the war was to show that in any case American output could seldom be predicted with an accuracy which made it possible for the United States government to make promises it could be sure of honouring. In 1942 the Americans were threatening to produce more tanks in 1943 than could conceivably be needed. In 1944 their output fell so far short of expectations that the British received only sixty per cent of their estimated allotment of Shermans and had reason to be thankful that they had rejected a proposal that they should reduce their own output. In general, where supplies were ample the British received what they expected; where they were scanty, what was available had to be shared out in accordance with *ad hoc* decisions. In 1943 lend-lease assignments provided about a quarter of the munitions furnished to British and Commonwealth forces, purchases in the United States about two-and-a-half per cent.[33] The corresponding figures for 1944 were 27·2 and 1·5 per cent. The United Kingdom provided about seven-tenths of the munitions furnished to British and Commonwealth forces during the whole of the war, the rest of the British Empire about a tenth, the United States about a fifth. Lend-lease assignments to the British accounted for roughly eleven per cent of the war expenditure of the United States, reciprocal aid to the United States for roughly nine per cent of Britain's.[34]

President Truman suggested in a report to Congress in 1945 that, insofar as a nation's contribution to the war could be measured in financial terms, probably the best measurement to take was the proportion of the national income devoted to its war effort. By that standard Britain was, of course, by far the largest contributor to the Allied cause. She spent a considerably higher proportion of her national income on the war than was spent by either the United States or the Soviet Union, although the Russians suffered a far heavier loss of life and played a major part in bringing the German Army to its knees. On the eve of the Allied landings in Normandy in 1944, fifty-five per cent of the employable population of the United Kingdom was serving in the armed forces or engaged in war production.[35] The corresponding figure for the United States was forty per cent. In other words, only forty-five per cent of Britain's labour force was available to provide the goods and services by which life was sustained, while in the United States fifty-eight per cent of the labour force was employed on tasks not related to the war effort and two per cent was unemployed.

However, although the British were more or less at full stretch from 1940 to the end of the war while the Americans always had a reserve of manpower to draw upon, there were many resemblances between the means by which the two nations equipped themselves for war. The British began to rearm (as a result of German rearmament) in 1935 and 1936; took a big step forward (as a result of Munich) in 1938; and threw themselves heart and soul into the business of war (as a result of Dunkirk) in 1940. The Americans began to refurbish their arms industry (as a result of orders received from the British and the French) in 1938 and 1939; took a big step forward (as a result of events in France and the Low Countries) in 1940; and went into high gear (as a result of Pearl Harbor) in 1941. Both countries started with a nucleus of government factories and specialist firms (although Vickers-Armstrongs had no exact counterpart in the United States); both depended for the growth of their productive capacity largely on contributions from thousands of firms, great and small, with little or no previous experience of the arms business. In both countries the supply departments encouraged the firms they dealt with to employ sub-contractors, and in some instances made it a condition of their contracts that they should do so.

But the parallel must not be pushed too far. American firms were, in general, larger than British firms. In the United States the final assembly of an aircraft in large-scale production might be done in a factory employing perhaps 20,000 to 40,000 workers; in Britain the number was more likely to be of the order of 3,000 to 15,000. On both sides of the Atlantic sub-contracting was common practice in peacetime, but very small sub-contracting firms, employing from ten to two hundred workers, were proportionally less numerous in the United States than in Britain. In the United States the federal authorities, finding that a migration of labour stimulated by rearmament tended to create distressed areas and ghost-towns, consciously adopted a policy of seeking out prospective contractors and sub-contractors in such places as a means of alleviating hardship. Largely as a result of loans and premiums offered to small firms to induce them to embark on war production, the dollar value of contracts let to firms with 500 or fewer workers rose from 12·6 per cent of the value of all contracts in 1943 to twenty per cent in 1944 and 28·5 per cent in the summer of 1945.[36] In Britain the supply departments favoured extension of the sub-contracting system as a means of enlarging and speeding up production rather than on social grounds. Once the rearmament drive was well under way, little seeking out was necessary. Main contractors and potential sub-contractors were equally alive to the advantages both could gain from coming together, and no special inducements were needed. By 1944 it would have been hard to find a metal-working or engineering firm in Britain which was *not* working directly or indirectly for the government.

No one knows precisely, or for that matter approximately, how many firms contributed to British war production, but an estimate, based on admittedly imperfect data, puts the total at probably well over 30,000. Even so, the larger firms were important. Hawker-Siddeley, and Vickers as proprietors of Supermarine and Vickers Aviation, were between them responsible for nearly half the military aircraft built in the United Kingdom during the war. Vickers were also, on the Vickers-Armstrongs side, large-scale producers of ships, tanks, guns and small arms. Imperial Chemical Industries became major producers of military explosives. The big motor firms built cars and lorries for the armed forces, and most of them also had a hand in aircraft production as sponsors or managers of shadow factories. Vauxhall Motors developed the Churchill heavy tank. BSA were the principal manufacturers of the Browning machine-gun, used in large numbers by the air force, and important producers of other machine-guns.

A feature common to British and American methods of procurement was that the role of such bodies as the Ministry of Production, the Office of Production Management and the War Production Board was confined to planning, co-ordination and adjustment. In the United States the War Department was responsible for the procurement of munitions and military aircraft; the Navy Department for that of naval vessels, naval aircraft and oil. The Maritime Commission was the procurement agency for merchant vessels; the Treasury dealt with metals, raw materials and certain manufactured goods. In Britain the government gave some thought, when the Ministry of Supply was first mooted, to making it responsible for procurement in general, but dropped the idea in face of objections from the Admiralty and the Air Ministry. The Admiralty retained its responsibility for meeting naval requirements, and early in 1940 enlarged its scope by taking over the Merchant Shipbuilding and Repairs Department of the Ministry of War Transport. The Ministry of Supply's task was more or less confined, therefore, to the procurement of material for the army.

The Air Ministry, and after it the Ministry of Aircraft Production, built their complex of suppliers almost entirely round the nucleus of the firms which had furnished the Royal Air Force with aircraft in time of peace. One or two additions were made towards the end of the war to the close circle of 'family firms', but demands for the creation of a national aircraft factory were resisted. The Royal Aircraft Establishment and the Aeroplane and Armament Experimental Establishment made valuable contributions to design, research and experiment, but not to production. At the same time the productive capacity of the specialist firms was enormously expanded by the hiving-off of additional factories, by sub-contracting and by the setting up of agency factories and production groups under expert guidance. In rare cases the Ministry of Aircraft Production took over the administration of a firm or encouraged a change of ownership, but in general the ministry was careful not to trespass on the province of directors and managers. Suggestions that manufacturers might use substitutes for scarce materials or standardize components were made with tact and even deference. The agency system was criticized in some quarters on the ground that payments to firms which undertook the management of agency factories were not high enough to promote efficiency, but the firms did not seem to mind being underpaid as long as they could feel they were doing a useful job. Towards the end of the war the Ministry of Aircraft Production had some eighty to ninety agency factories in operation and could point to a threefold increase in production since the outbreak of war as at least a partial justification of its methods.

When the size and weight of the aircraft produced are taken into account, the increase becomes even more impressive. In 1939 no heavy bombers were completed, and more than eighty per cent of the aircraft delivered were trainers, fighters or light bombers.[37] Of the 24,461 aircraft delivered in 1944, 5,507 were heavy bombers. These figures do not include the 4,052 aircraft produced in Canada in 1944 either on British or on Canadian account. The Americans produced 93,623 military aircraft in 1944, the Germans 39,275 aircraft of all types.[38]

The Admiralty continued during the war its peacetime practice of relying mainly on private enterprise for the navy's needs but making a limited use of naval dockyards and factories. Of some 668,000 people employed on Admiralty work at the end of the war in Europe, about 36,000 were working in the Royal Dockyards and some 70,000 in other Admiralty establishments.[39] Nineteen agency factories were working on Admiralty account.

The Ministry of Supply differed from the other procurement agencies in making more use of government factories. This tendency arose partly from a feeling that shell-filling was a process for which commercial firms were not particularly suitable, partly from something akin to necessity. When rearmament began the number of firms with recent experience of arms manufacture which were free to devote more than a small part of their resources to army contracts was very small. New factories were essential, and there was an obvious limit to the extent to which a handful of specialist firms could be called upon to supervise their construction and equipment without detriment to their other activities. On the other hand there was a considerable fund of accumulated knowledge of such work to be drawn upon at the existing government factories at Woolwich, Enfield and Waltham Abbey. It was therefore natural that new ordnance factories, some of them very large, should be formed in the image of those already in existence. In this way forty-three Royal Ordnance Factories came into being between the beginning of the rearmament period and the end of the war. Controlled by the Ministry of Supply and managed by civil servants, they were expected to pay their way and to compete on level terms with private enterprise. It was, however, always the Ministry's intention that the biggest contribution to wartime production should be made by the commercial engineering industry. Prospective main contractors were given 'educational' or trial orders, sub-contracting was adopted on an increasing scale, and ultimately practically the whole of the engineering and metalworking industries, and a great many firms outside those industries, were drawn into war production. Many firms worked not only for the Ministry of Supply but also for the Admiralty, and sometimes for the Ministry of Aircraft Production as well. The Ministry of Supply also made extensive use of agency factories, employing no less than 159 by 1945. In broad terms, about a quarter of the munitions produced in the United Kingdom during the war and the rearmament period came from government factories of one kind or another.

The Germans, like the Americans and the British, relied chiefly on private enterprise to equip their armed forces, but it was private enterprise with a difference. The importation of foreign workers, the illegal employment of prisoners of war in arms factories and recourse to slave labour enabled Germany to maintain a huge output of munitions despite heavy bombing of her centres of population and without fully mobilizing her womanpower as Britain was obliged to do. The process began soon after the occupation of Prague with the arrival of Czechoslovakian workers. These were described as volunteers, received comfortable billets and were treated reasonably well, at any rate in the early stages. Later arrivals included French, Belgian and Dutch workmen who were also deemed to be volunteers. When Polish and Russian prisoners of war and deportees began to arrive in large numbers in 1941 they were at first segregated from the earlier arrivals, but later ethnic distinctions became blurred and workers from Western Europe who had come to Germany more or less voluntarily found themselves living behind barbed wire. About the same time French, Belgian and Dutch prisoners of war were drafted into arms factories. Albert Schrödter, manager of the Krupp-owned Germaniawerft, is said to have pointed out to Alfried Krupp that the employment of prisoners of war for arms production was contrary to international law, and to have been told that they were already being employed at Essen. In general, such objections were brushed aside by the authorities on the ground that the Soviet Union was not a party to the Hague Convention and that the customs and usages of war did not apply to Russians. There is no doubt that the firm of Krupp was deeply implicated in the illegal employment of prisoners-of-war, their appalling ill treatment, and the use of slave labour for the construction of the Berthawerk arms factory in Silesia, but responsibility for these crimes proved hard to fix. In 1941 Gustav Krupp suffered the first of a number of strokes, and in the following year Alfried took his place as executive head of the family business. Alfried said after the war that he was not the creator of the system he inherited, did not fully understand it and could not be assumed to have approved of it. Co-operation with the National Socialist government, he said, was not a crime: for a German it was a patriotic duty.

On 15 November 1943, Bertha Krupp formally renounced her ownership of the Krupp concern in favour of Alfried. The designation of Alfried as sole heir contravened a statute of 1920 which decreed that only an only child could be so designated, but was retrospectively sanctioned with Hitler's approval. Some eighteen months later American troops under the strategic control of the British Twenty-First Army Group entered Essen. Since Essen was within the British occupation zone, they were relieved in June by British troops. Pending the formal seizure of the firm's assets by the British Military Government in the following November, the Krupp works were allowed to make steel roofing, start an assembly line for locomotives, and draw up a reconstruction programme which would entail, among other tasks, the building of a new railway station.

THIRTEEN

World Without Peace

In the autumn of 1944 the British government made some anxious calculations about the country's economic future and the probable effect on the economy of the cessation of hostilities. In the interests of the national war effort production for civilian consumption had been reduced to half the normal peacetime level. Imports had been drastically reduced. Export markets had been lost, and foreign investments to the value of a thousand million pounds had been liquidated.[1] The national gold and dollar reserves, after falling almost to zero in 1941, had risen to some £420 million as a result of spending by and on behalf of American troops quartered in Britain, but were expected to decline to some £250 million by the end of the war if nothing were done in the meantime to put matters right.[2] External liabilities, incurred largely as a result of the campaigns in the Middle East, amounted to some £3,000 million, and this figure did not include any sum that might have to be paid to the United States government for lend-lease material.

At that time the war with Germany was expected to be brought to a successful conclusion about the end of 1944, but the war with Japan to continue for some eighteen months after the Germans were defeated. The government concluded that a start must be made during those eighteen months with preparations for a return to peacetime conditions. Production for civilian consumption must be increased, if possible by fifty per cent, but this must be done without detriment to the contribution the government intended to make to the war in the Far East and the Pacific. Some expenditure on housing, public utilities and the replacement of worn-out plant and machinery in industries catering for civilian and export markets must be undertaken. Foreign trade must be revived, additional workers must somehow be found for industries—such as coal-mining and the textile industries—whose output had declined. External liabilities must be partly offset by improved trade balances and larger reserves of gold and foreign currencies.

At the same time it was, the government felt, essential that these things should not be done at the cost of imperilling the country's future relations with the United States. Equality of sacrifice, the pooling of resources implied consultation. At a series of meetings in October and November the Americans agreed, in the light of a frank disclosure of Britain's problems, to continue to meet her lend-lease requirements after the defeat of Germany, on the understanding that these would include not only munitions but also food, oil and shipping.[3] A modest increase in Britain's gold and dollar reserves would not be discouraged, and no objection would be made to attempts she proposed to make in 1945 and later to improve her foreign trade. The Americans were not prepared to sign a protocol which would give the British the same assurance as was extended to the Russians that their requirements would be met, but that was hardly to be expected. They recognized that Britain had done everything it was in her power to do for the common cause and that, short of inviting national bankruptcy, she would continue to do everything she could to bring about the collapse of Japan when the war with Germany was over.

However, Allied estimates of the time needed to subdue the Axis powers proved seriously at fault. The war with Germany did not cease at the end of 1944, but continued until the following May. The Japanese did not hold out for eighteen months after the Germans were defeated, but surrendered in August 1945. In the meantime the Soviet government's imposition of Communist rule on Poland and Rumania made it all too clear to the Western Allies that the cessation of hostilities would not bring them a secure peace.

The illness and death of President Roosevelt in the spring of 1945 did not make things easier. Roosevelt was already a sick man when he made far-reaching concessions to the Russians at the Yalta conference. For some weeks before his death in April he was unable to do a full day's work. When the British raised at the beginning of the month a politico-strategic question of the utmost importance, the issue was referred to a subordinate who did not feel competent to pronounce upon it. The decision to drop two nuclear bombs on Japan was made, after consultation with the British, by a recently-installed President who had never, as Vice-President, been admitted to Roosevelt's full confidence.

A few days after the Japanese surrender, President Truman issued a directive ending lend-lease with effect from a date not yet fixed, but which turned out to be 2 September 1945. Since failure to receive supplies already in the pipeline on that date would have completely disrupted Britain's economy, the British government agreed to purchase them on terms to be negotiated in the context of a general settlement. The cost of these supplies brought the total value of lend-lease material furnished to Britain since 1941, less the value of reciprocal benefits, to roughly $21,000 million. The British were called upon to pay only $650 million, made up approximately as follows:

	$
Supplies delivered after VJ-day	118,000,000
Payment for fixed assets valued originally at $350,000,000	60,000,000
Payment for stocks held by UK on day of settlement	472,000,000
	650,000,000

No other claims were made. No charge was made by either side for material consumed or destroyed during the war. Surviving naval or merchant vessels were returned to the countries that had supplied them. Installations constructed under reciprocal aid for United States forces in the United Kingdom reverted to the British, and *vice versa*.

Unfortunately other legacies of the war could not be wiped off the slate so easily. Britain and the United States were constrained by fear of the Soviet Union not only to maintain large and exceptionally well equipped forces in their occupation zones in Europe, but also to add substantial numbers of nuclear weapons to their stockpiles. The partial or complete disruption of the British, French, Dutch, Belgian and Portuguese colonial empires created a whole host of new sovereign states, each of which claimed the right to maintain armed forces as an outward and visible sign of sovereignty. At first most of them depended on their former suzerains for weapons and other military equipment, but before long some of them began to look elsewhere.

Furthermore, within a few years of the end of World War II political, ideological and economic conflicts in Europe and Asia launched the arms trade on a boom which has continued to this day. Whether that is reckoned a good or a bad thing must depend, of course, upon the point of view. The Berlin air lift—undertaken by the Western Allies when, in 1948, the Russians cut land communications between Berlin and West Germany—has been described as saving the American aerospace industry from bankruptcy. Although that may not be literally true, there is no denying that many thousands of workers and shareholders in firms concerned with the manufacture of military aircraft and ancillary equipment have some reason to be grateful for the increased tension between the Soviet Union and the Western Powers which accompanied and followed the Berlin blockade and the Korean War of 1950–1953. Moreover, even the most convinced advocate of disarmament must admit that the piling up of armaments by Russians and Westerners since 1945 has not, as yet, had the dire consequences that might have been expected.

The fact remains that the leading capitalist powers and the Soviet Union are still at loggerheads, and seem likely to remain so. The relation that subsists between them can be best described as an uneasy truce arising from mutual reluctance to precipitate a nuclear holocaust. The balance of power at which European statesmen used to aim has been replaced by a balance of terror. In any case the good fortune which has enabled us, up to the present, to escape a third world war should not blind us to the fact that, from the point of view of a citizen of the world, the doors of the temple of Janus have seldom been closed for more than a few days at a time during the past thirty years and more. The author of a book published in 1969 calculated that there had been, up to that date, fifty-five wars of 'significant size, duration and intensity' since 1945.[4] He added that the number would rise to more than three hundred if *coups d'état*, large-scale riots and 'clashes of unorganized, low-order violence' were included. And violent affrays of either a low or a high order have certainly not grown less frequent since 1969.

In some of these conflicts staggering quantities of war material have been consumed. The tonnage of bombs used in the Korean War exceeded the tonnage dropped by the Western Allies in the Pacific theatre throughout the Far Eastern war of 1941 to 1945. A greater weight of bombs was dropped on Vietnam between 1965 and 1967 than the Allies dropped on Germany during the whole of the Second World War.

Warmaking on such a scale must have brought much profitable business to arms manufacturers. Whether arms manufacturers can be justly accused of fomenting conflicts between capitalist and communist states in their own interests or the interests of their shareholders is another matter. A certain amount of lobbying by hawks whose hawkishness may not always be inspired by undiluted idealism can be assumed, but its effect in any particular instance is hard to judge. What we do know is that governments, besides committing their own countries or vassal countries to external wars, have on quite a number of occasions since 1945 precipitated uprisings in foreign countries by allowing their agents to encourage dissident minorities to rebel and by conniving at the acquisition by such minorities of arms and ammunition. This has happened not only in Cuba but also in the Middle and Far East and in Africa.

The question of the extent to which governments ostensibly dedicated to the preservation of law, order and international harmony are to blame for the manifest increase of lawlessness throughout the world since the end of World War II is complicated by the fact that the governments of some capitalist countries have become in comparatively recent years not merely politically and economically but financially interested in the sale of arms to foreign buyers. (In communist countries, of course, the state has always had a monopoly of such transactions.) Their incursion into the distributive side of the arms trade is commonly attributed to the tendency of weapons and weapons-systems to become more complex and more expensive. Instruments of aggression or national defence, it is often said, have become so sophisticated that only organizations backed by the resources of the state can develop and market them.

That, however, is true only of a limited number of weapons, some of which (such as nuclear bombs and warheads) are not articles of commerce. Articles classed as defence equipment which capitalist governments sell or help to sell include guns, tanks, warships, naval and military aircraft, automatic weapons and small arms, to say nothing of a variety of ancillary equipment ranging from earth-movers and complex electronic devices to boots and buttons. Some of these are made in government factories; many more produced by commercial undertakings which may or may not belong wholly or partly to the state.

What actually happened at the end of World War II was that the armed forces of the leading belligerent powers found themselves with large stocks of surplus weapons. In the ordinary way they would have stored some of these and sold others to authorized dealers in such material. To some extent they did both. Before long, however, the United States government began to buy from its own armed forces surplus material which it proceeded to bestow on allies or potential allies in Europe and elsewhere. Recipients included not only a number of European and Latin-American countries but also the governments of South Korea, the Philippines, Taiwan and some Middle Eastern countries. The countries concerned were thus endowed with the means of resisting communist aggression, while the armed forces of the United States received funds which they could spend on the acquisition or development of new weapons without having to seek direct approval from Congress.

About the middle of the 1950s, the United States government began to furnish members of the North Atlantic Treaty Organization not merely with weapons left over from World War II but with defence material of recent design. This was supplied free of charge to the poorer members of the organization, but countries which could afford to do so were expected to manufacture the appropriate weapons in their own factories and to pay royalties to American firms under licensing agreements. An organization under a civil servant named Henry Kuss was set up by the military authorities in 1957 to supervise these arrangements.

This was the first step towards direct sales. By 1956 the high cost in foreign exchange of the maintenance of American forces abroad was becoming a burden on the Treasury. Royalties paid to firms in the United States provided welcome relief.

The next step was to persuade foreign governments not merely to permit the manufacture of American-designed weapons under licence, but to buy American weapons manufactured in the United States. In 1961 the Secretary of Defense, Robert S.

McNamara, set up the Office of International Logistics Negotiations (afterwards renamed) for the purpose of selling defence material to foreign governments. The new organization was provided with offices in the Pentagon building in Washington and Kuss was put in charge of it. A Red Team was appointed to sell arms to Canada, the Far Eastern countries, Scandinavia, France, and a number of countries associated with the North Atlantic Treaty Organization; a White Team to conduct negotiations with the West German Federal Republic; a Blue Team to deal with sales to Latin-American countries, Italy, Spain and the Benelux countries; and a Gray Team to negotiate with the United Kingdom, Switzerland, Austria and the Near and Middle East. By persuading some of the 1,400 to 1,500 American firms which ranked as suppliers of defence equipment to interest themselves in export markets, and by conducting a series of vigorous sales campaigns on their behalf, Kuss greatly increased the proportion of war material sold to war material given away. The federal government continued, however, to be by far the biggest purchaser of American-made arms.

In Britain the service departments were expected for some years after World War II to dispose of their own surplus material in consultation with the Ministry of Supply. As early as 1955, however, the staff of the British Embassy in Bonn included a Supply Attaché who concerned himself to some extent with the sale of weapons and weapons-systems. Three years later the Ministry of Supply set up a section of the staff, under an assistant controller, for the purpose of promoting foreign sales of defence material, and if possible (which it was not) persuading members of the North Atlantic Treaty Organization to adopt common standards. About the same time, service attachés at all appropriate embassies were asked to pay more attention to attempts to promote sales.

Despite these efforts, the monetary value of British exports of arms fell sharply in the late 1950s and early 1960s. This was doubly disappointing because the British knew that they had, in the Chieftain tank and the British Aircraft Corporation's Lightning fighter, two weapons of outstanding quality which ought not to be hard to sell.

Part of the trouble was that few service attachés were equipped by temperament or training for the role of salesman. Recognizing this, in 1965 the Labour government then in office asked Sir Donald Stokes, Managing Director of the Leyland Motor Corporation, to advise the Ministry of Defence and the Ministry of Aviation on the promotion of foreign sales. The sequel was the appointment in the following year of Raymond (afterwards Sir Raymond) Brown to a post which made him, in effect, chief government arms salesman.

Drawing on his experience as Chairman of the Racal group of companies, which had achieved considerable success as exporters of electronic equipment for both civil and warlike purposes, Brown set up an organization staffed mainly by civil servants, supplemented by a few serving officers of the armed forces. Sales of material for naval and military purposes were dealt with by separate divisions within the Ministry of Defence. With the Exports and International Relations Division of the Ministry of Technology (which absorbed the Ministry of Aviation in 1966), these divisions formed a complex known collectively as the Defence Sales Organization. By 1970 the organization had a staff of nearly three hundred, of whom about sixteen were more or less permanently stationed abroad.

In 1969 Sir Raymond Brown was succeeded by Lester (afterwards Sir Lester) Suffield, a sales director seconded (initially for two years) from the Leyland Motor Corporation. The army and navy divisions were then replaced by divisions organized on a geographical basis. Other changes sponsored by Sir Lester Suffield were designed to ensure that customers received prompt delivery and efficient after-sales service. He and his successor must, however, have found the first of these objectives rather hard to attain in the troubled conditions of the 1970s.

Between 1967 and 1972 Sir Raymond Brown, Sir Lester Suffield and their staffs succeeded in increasing British exports of defence material by some eighty per cent. This they did chiefly by furnishing commercial firms with introductions and recommendations, by sponsoring and sometimes financing demonstrations of defence equipment made by commercial firms or in government factories and (from 1969) by circulating annual catalogues of the material available.

In France there is a long tradition of close co-operation between governmental or quasi-governmental agencies and commercial firms with a substantial stake in export markets (a fairly high proportion of which are in any case owned wholly or partly by the state). In 1961 a body called the Délégation Ministérielle pour l'Armament (DMA) was set up to supervise at a high level the production and sale of defence equipment, and was made directly responsible to the Ministry of Defence. Some four years later its export department was expanded to form an organization called the Direction des Affaires Internationales (DAI), whose functions are in some respects not unlike those of the British Defence Sales Organization. DAI despatches market-research missions and organizes demonstrations and displays of defence equipment furnished at its prompting by manufacturers. Other organizations with government backing which help to promote foreign sales include the Office Français d'Exportation de Matériel Aéronautique (created in 1950); the Office Général de l'Air (an association of aircraft manufacturers whose origins go back to 1937); the Société Française d'Exportation d'Armaments Navals; and the Société Française de Matériel d'Armaments. The last two are bodies set up with government backing to do for manufacturers of naval and military equipment respectively what the Office Française d'Exportation de Matériel Aéronautique (OFEMA) and the Office Général de l'Air (OGA) have done for manufacturers of aircraft.

An advantage claimed for this system is that it enables manufacturers to dispense with large sales organizations and concentrate on research, development and production. The existence of two organizations concerned with foreign sales of aircraft might be expected to lead to wasteful overlapping, but apparently this is avoided by an agreement between OFEMA and OGA which gives each of them its own sphere of interest.

The Soviet Union began soon after World War II to distribute arms to countries with which she was linked by ideological ties. Her first post-war arms deals with non-communist countries appear to date from 1954, when she sold arms to Syria and Guatemala. Thereafter she became a willing purveyor of war material to customers of almost any political complexion who were prepared to do business with her.

For the Soviet Union the promotion of foreign sales of arms by government agencies is, of course, no novelty. Even before World War II her arms deals with foreign buyers may not always have been prompted solely by ideological or strategic considerations, although doubtless such motives were always present. During the undeclared war of 1937 to 1945 between Japan and Nationalist China, Chiang Kai-shek received large quantities of war material from the Soviet Union on extremely favourable terms, although he was also receiving supplies from Britain and the United States. The Spanish Republican government, on the other hand, paid heavily in gold for the military aid it received from Russia during the Spanish Civil War. There is no doubt that in recent years the Soviet authorities have been fully alive to the economic benefits they derive from the export of arms.

In a sense the practice of the capitalist world with respect to foreign sales of defence equipment can be said, therefore, to have moved closer to that of the communist world. There is still the difference that, while all arms sold by communist countries are made in government factories, most of those sold by capitalist countries are produced by joint stock companies or corporations. The difference is not, however, always as great as it may seem. In the United States a traditional distaste for 'state socialism' does not prevent government and Congress from keeping a sharp eye on multi-million-dollar corporations which do business with the federal authorities. The British Defence Sales Organization promotes the sale of defence equipment by commercial firms, but it also sells weapons manufactured in government factories. In France all defence equipment comes from factories organized on a commercial basis, but some of the most important of these factories belong to the state, others to firms closely linked with the state. A number of capitalist countries, such as Italy, the Netherlands, Sweden and Switzerland, grant virtual monopolies of the manufacture of certain defence equipment to quasi-nationalized concerns.

In terms of revenue and foreign exchange earned, government participation in the promotion of foreign sales of arms has worked well for the Americans, the British and the French. A criticism sometimes made of the system is that it could lead to a conflict between national interests and the requirements of export markets. In Britain, for example, the Defence Sales Organization is represented on committees which, in effect, decide what weapons are to be developed for future use. Does this mean that Britain's armed forces could one day be saddled with weapons less suited to their needs than to those of customers abroad?

The answer is, of course, that the advent of government arms salesmen has merely given a new twist to an old problem. There must always be a risk that weapons developed to suit the needs of the country of origin may be hard to sell to other countries. They may have too limited a function. They may be too sophisticated, or too expensive, for foreign buyers. They may lean too heavily on ancillary equipment too secret to be exported. Conversely, a weapon developed with too keen an eye to export markets may, for a variety of reasons, fall short of domestic needs.

A number of ways of circumventing this difficulty have been tried. The traditional solution is to sell redundant or obsolescent equipment to foreign buyers at bargain prices. The Russians are reputed to have sacrificed financial to economic and politico-strategic benefits by supplying a number of North African, Middle Eastern and Asiatic countries with MiG-21 fighters at prices well below cost. Doubtless there will always be markets somewhere in the world for surplus or outmoded weapons, but large-scale transactions of this kind have become increasingly difficult to negotiate in recent years, because massive rises in the prices of raw materials have tended to make once backward and impoverished countries more affluent and more exacting.

Where ample funds are available for research and development, a weapon expressly designed for overseas markets may provide the answer. The F-5 Freedom Fighter was developed by Northrop, with financial backing from the United States government, purely for export. Although its adoption by something like a dozen countries justified the faith shown in it by the American authorities, probably few governments could afford to make such experiments very often.

A less radical solution is to develop weapons which serve more than one purpose or can be readily modified to meet various requirements. The Dassault Mirage falls into this category, as do a number of multi-role aircraft manufactured or projected by British or Continental firms, or by combinations of such firms. American aircraft which can be classed as multi-purpose include the McDonnell Phantom II and—despite its name—the much-criticized Lockheed F-104 Starfighter.

An expedient favoured by the British is to offer the prospective purchaser a custom-built weapons system, consisting of standard weapons or modifications of standard weapons, with ancillary equipment more or less tailored to his requirements. Britain's much-discussed 'package deal' with Saudi Arabia in the 1960s was a notable example of the application of this method.

As a result of the wooing of Arab chieftains by the British and British Indian governments before and during World War I, most of the Arab states which came into existence after the collapse of the Ottoman Empire looked between the wars to Britain to meet their defence requirements. Saudi Arabia, in particular, bought obsolete or surplus British aircraft to patrol her large and sparsely-populated territory. After World War II she acquired a number of American aircraft, but her air force proved incapable of preventing violations of her air space when it was put to the test by the eruption of a civil war in the Yemen. By 1964 she was known to be contemplating the purchase of supersonic fighters, and was thought likely to choose either the Lockheed F-104 Starfighter or the Northrop F-5 Freedom Fighter. The British Aircraft Corporation and the French firm of Dassault were also in the running, but were not regarded in 1964 as very serious competitors.

On-the-spot enquiries led, however, to the conclusion that the mere acquisition of new fighter aircraft would not meet the country's needs. Mr Geoffrey Edwards, a British business man who had gone to Saudi Arabia in search of building contracts, was one of a number of intermediaries who formed the opinion that what the Saudi Arabians really wanted was not just a few squadrons of modern fighters but a comprehensive air defence system, complete with early-warning radar stations and an up-to-date communications network.

It happened that at that stage the British government of the day suspended development of the British Aircraft Corporation's TSR-2 strike aircraft and took an option on the General Dynamic Corporation's F-111A, then just coming into production in the United States. The F-111A promised to be cheaper in the long run than the TSR-2, but dollars would be needed to finance the transaction. To enable the British to recover part of the cost of the F-111As covered by their option, the United States government agreed that British firms should be allowed to compete on level terms with American firms for up to $325,000,000-worth of American defence contracts, and that American and British firms should make joint attempts to sell $400,000,000-worth of defence equipment to third parties.

The option was due to expire on 1 March 1966. In 1965 the British concluded with Saudi Arabia an agreement by which the Saudi Arabians undertook to buy some $275,000,000-worth of British fighters, trainers and early-warning and signals equipment, in addition to a substantial number of American-made Hawk missiles. Pending delivery of the Hawk missiles, the Saudi Arabians afterwards placed an order for British-made Thunderbird missiles to the value of some $20,000,000.

The Americans were gratified by the sale of the Hawk missiles, but not so pleased when the British decided not to take up their option on the F-111A. However, the Saudi Arabian armoury now includes American as well as British aircraft, in addition to British and French armoured cars and American and French missiles. Further purchases of American aircraft are planned.

British and American spokesmen afterwards gave different accounts of the circumstances in which the package deal was negotiated. According to the British, the Americans gave them a clear field.[5] Henry Kuss denied that he went so far as to withdraw his offer to the Saudi Arabians of the Lockheed F-104 or the Northrop F-5. He said that he merely told the British that he would not object to their including Hawk missiles in any package they might offer.[6] But he can scarcely have failed to know that any such package would include BAC Lightning fighters, so the difference between the two versions is perhaps less great than it may seem.

Many examples could be cited of 'offset' deals comparable with the arrangement by which Britain was to have financed her acquisition of the F-111A. Within the past ten or twelve years Belgium has recovered all or most of the cost of tanks bought from the Germans by rendering them services and selling them goods, including defence equipment, which they would not have bought in the normal course of trade. She has defrayed the greater part of the cost of a large batch of Mirage Vs by entering into a licensing agreement with Dassault. The Netherlands and Norway have bought German defence equipment on exceptionally favourable offset terms, and Brazil is said to have offered to pay for Mirage Vs in coffee and lobsters.

At this point it must be emphasized that the export of arms from the leading industrial countries except under licence is still forbidden, as it has been for the past forty years and more.

Despite the advent of government arms salesmen and the creation of such supra-national institutions as the United Nations and the European Economic Community, any decision to export arms from such a country is still essentially a political decision. It is a decision the government of a sovereign state must make in the light of its conception of the national interest, of treaty obligations, of any representations made to it through diplomatic channels or the United Nations, of the opinions and prejudices of its own members and supporters. If the present British government, for example, decides not merely in principle but in practice to sell Harrier aircraft to China despite protests from the Soviet Union and its own left wing—as it seems while these words are being written to be about to do—then that decision will be made not by the Defence Sales Organization but by the Cabinet. The Defence Sales Organization sells arms or promotes their sale only where it is allowed by its political masters to do so.

This does not mean, of course, that the details of every transaction are scrutinized at a high level. The procedure for granting export licences differs from country to country. Even so, it is a fair guess that the government of every country with an export trade in arms maintains, and from time to time revises, a list of countries whose requests for defence equipment will normally lead to the granting of licences more or less as a matter of routine; a list of countries not to be supplied with arms in any circumstances during the lifetime of the existing government (and therefore unlikely to ask for arms unless or until a more amenable government comes to office); and a list of 'borderline' countries whose requests must be referred for discussion at the ministerial level before the granting of licences can be considered.

It may well be wondered why such safeguards don't seem to prevent the means of waging war or launching rebellions from repeatedly getting into the wrong hands. However, to ask that invites the response that there is no general agreement as to which are the wrong hands and which the right ones. Let us begin, then, by putting the question in this form: If the arms-producing countries could agree (which at present they can't) to permit the sale of arms only to customers of whom all approved, would the licensing system be effective?

The answer is that probably it would work fairly well in the case of some weapons, but not where others were concerned.

Enormous numbers of small arms are held by dealers in surplus and used weapons or other non-governmental agencies in many countries. Most dealers in surplus and used weapons are law-abiding, but some are not. The chances are that it will always be possible to buy rifles, carbines, pistols and sub-machine guns in parts of the world where export controls are virtually non-existent, or are not enforced. Even where they are so strictly and so efficiently enforced that the smuggling of illicit cargoes under cover of darkness becomes almost impossible, there are various ways in which they can be evaded. Cases have come to light in which consignments of arms were cleared at the port of exit on the strength of documents either forged, or signed by officials not authorized to sign them.

Alternatively, the documents may be genuine, but obtained by misrepresentation. In the autumn of 1966 an aircraft left a Dutch airport with a cargo of Thompson sub-machine guns consigned to a respectable firm in Britain for reconditioning and return to the Netherlands. The pilot showed import and export licences which the firm had obtained from the British authorities in the belief that a genuine transaction with a customer for whom it had reconditioned weapons in the past was in prospect. On reaching the neighbourhood of the British airport which was his stated destination, he reported by radio that the owners of the aircraft had ordered him to divert it to Majorca. He then made for Biafra by way of Majorca, Algeria and the Sahara, but lost his bearings and almost ran out of fuel. Ultimately he made a crash-landing in West Africa, with the result that his true objective was revealed.

Large weapons such as tanks and military aircraft are not as easily smuggled as rifles or Tommy-guns. This does not mean that governments cannot acquire them against the wishes of the country of origin. A country which buys American or British aircraft may be required to sign a declaration to the effect that they are intended solely for its own use and will not be resold, but such undertakings are not enforceable. At the cost of a little passing unpleasantness, the purchaser can easily hand them over to a third party either in open defiance of his obligations, or on the pretext that they have to be sent abroad for overhaul or repair. Iran sent a large batch of American-designed but Canadian-built F-86 Sabres to Pakistan in 1966 on just such a pretext. Almost simultaneously, India received from a West German dealer in surplus weapons a number of aircraft ostensibly consigned to Italy.

The fact remains that the most potent single cause of the proliferation and wide dissemination of lethal weapons in recent years has been the cold war between the Soviet Union and the leading capitalist countries. Its effects have been seen not only in Europe and in the Far East—where attempts to halt the march of Communism by force of arms have led to untold suffering and destruction, to say nothing of humiliating setbacks for the Western powers—but also in the Middle East and Africa. Until about the middle of the 1950s Britain, France and the United States had a virtual monopoly of the supply of arms to Middle Eastern countries. Soviet arms deals with Syria and Egypt in 1954 and 1955 ended that. The sequel was the Suez affair, followed by lasting tension and bitter conflicts between Israel and Egypt. More recently, the Soviet Union has gained ever-increasing power, influence and strategic advantage in Africa by exploiting anti-colonial or anti-capitalist sentiment and by arming and otherwise supporting governments or factions willing to go along with her.

Little or no effective opposition has been offered to these incursions. Communist penetration of African countries has been helped rather than hindered by gestures intended to demonstrate international solidarity with victims or supposed victims of white supremacy or capitalist exploitation. In 1963, when the future of Angola was in question, sixty-two nations agreed not to supply Portugal with arms for the purpose of enabling her to retain her colonial possessions. In view of the obvious disadvantages of a premature or ill-prepared Portuguese withdrawal from Angola, no doubt some of them took comfort in the reflection that Portugal was already fairly well provided with British-built warships and aircraft of American or German origin. Moreover, as a member of the North Atlantic Treaty Organization she could scarcely be denied the means of keeping her armoury up to date. She was, however, asked to give an assurance that weapons sold to her in future would not be used in Africa. Anyone who believed that such declarations of intent would do anything to secure for Angola a smooth and bloodless transition from colonialism to independence was tragically mistaken.

In general, attempts on the part of the family of nations to regulate African affairs by means of arms embargoes have been conspicuously unsuccessful. The Security Council of the United Nations—again in 1963—invited all member-states to resist the South African policy of *apartheid* by discontinuing the supply of arms, ammunition and military vehicles to South Africa. In the following year the ban was extended to plant and materials suitable for the manufacture of defence equipment. But South Africa already had all the weapons, plant and materials her government was likely to need to enforce *apartheid*, if not in 1963 at any rate soon afterwards. In 1965 the South African Minister of Defence made a statement to the effect that no fewer than 127 licences for the manufacture in South Africa of defence equipment of foreign design were in existence.[7]

In that instance the resolution of the Security Council would seem to have been not merely ineffective, but misconceived. In substance the object in view was to convince the South African authorities that, although for the time being there was no general agreement among members of the United Nations to excommunicate South Africa, or to sever trade relations with her except so far as armaments were concerned, the policy of *apartheid* was so widely condemned that they might do well to consider whether they could afford to persist in it. A futile arms embargo was not likely to increase the chances that representations to that effect would receive a patient hearing.

In the case of Rhodesia, attempts have been made not only by means of an arms embargo but also in other ways to punish her for her defiance of British suzerainty and her long adherence—now abandoned but not forgiven—to the principle of white supremacy. Within a fortnight of her unilateral declaration of independence in 1965, the United Nations called for a ban on the sale of arms to Rhodesia.[8] Britain took the lead, as she was bound to do, in imposing a general ban on trade with her errant offspring. But Rhodesia could no more be deprived of arms than she could be deprived of oil. In 1967 the Secretary of the Rhodesian Treasury testified that the country was receiving adequate supplies of arms, ammunition, aircraft and military vehicles, and of equipment and materials needed for their maintenance or, in appropriate cases, for their manufacture.[9] Some twelve years later, the position is that the Rhodesians have gone a long way to put their house in order, but have not yet been able to comply with the insistence of the British government that they should come to terms with minority leaders who seek to demonstrate their fitness for office by sponsoring acts of terrorism in which weapons of Soviet origin have been used.

Does this mean that arms embargoes or more comprehensive economic sanctions should never be tried? Clearly not. The League of Nations, for all its shortcomings, did succeed by such means in averting or curtailing a number of minor conflicts during its relatively brief existence. We now know that economic sanctions, if extended to oil, could have forced Mussolini to call off his calamitous Abyssinian venture. Circumstances in which such measures could be effective may arise in future. Where public opinion calls insistently for the withholding of supplies on moral grounds, an elected government may perhaps be justified in deferring to it, even though the chances of success appear slender. But a government which does that ought surely to take all possible steps to ensure that the public knows the facts. Successive British governments during the past twelve years and more cannot have been right to allow people to believe that an effective ban on the supply of oil to Rhodesia was in force when ministers were in a position to inform themselves that it was not.

Whether Rhodesia achieves true independence under a stable government or becomes yet another focus of intrigue and dissension must be a matter of grave concern to the Western world. Events in the Horn of Africa touch Americans and Europeans more closely than many of them know. Israel and Egypt may or may not agree to sink their differences. Iran has been shaken by upheavals which cannot have failed to make her northern frontier more vulnerable, yet her new leaders propose to reduce her intake of armaments. For the moment, however, all these events and preoccupations are dwarfed by the problem of China.

China is not in a position to sustain either an offensive or a defensive war with the Soviet Union. She hopes to become so with help from the West. Her requests for arms are not likely to stop at Harrier aircraft. She needs tanks, guns, and much else besides to modernize an army at present equipped largely with old and obsolescent weapons. She also needs, and intends, to increase her own arms-producing capacity by expanding her heavy industry, again with help from the Western powers. She is rich in reserves of non-ferrous metals, so she will be able to pay for her imports if the plant and technical assistance needed to develop these resources are forthcoming. She is also rich in coal. In view of her present shortcomings, it seems likely that her spokesmen are sincere when they assert that her operations in Vietnam are of a strictly limited character and will not be pushed to a point which entails the risk of a serious confrontation with the Soviet Union.

Fifteen or twenty years ago there were people in Britain—some of them in highly responsible positions—who said that the Western powers did not really need to worry about the Russian problem, because the Chinese would solve it for them. They argued that the Chinese and the Russians were bound eventually to come to blows. Britain and the United States would then be able to watch the struggle from the sidelines.

Probably there is now no one in the United Kingdom who thinks along those lines. We recognize that war between China and the Soviet Union would be a disaster. Moreover, the idea of using China as a pawn—or even as a bishop or a knight—in the long-drawn chess-game between the Western powers and the Soviet Union has become repugnant. Even so, there is a respectable case for helping to arm China to the point at which she could resist a Soviet onslaught. In military matters the Russians have shown themselves during the past thirty years to be realists, unwilling to bite off more than they can chew, willing to draw back when they see that success is impossible or would be bought too dearly. The inference is that they would not commit themselves to an all-out war with China unless they felt reasonably sure of winning it. If China were strong they could feel no such assurance. Unhappily the ideal of a peacefully-inclined world with no weapons except policemen's truncheons is unattainable. Since we have to live in one which bristles with missile-projectors and buzzes with military aircraft, a balance of forces would seem to offer the best chance of security.

Notes and Sources

1 Merchants of Death?

1 Forde, *Habitat, Economy and Society*, 298
2 Vickers Limited, *Annual Report and Accounts*
3 Scott, *Vickers*, 240. See also Chapter 9

2 Beginnings

1 Beeler, *Warfare in Feudal Europe*, 37
2 Hogg, *Artillery*, gives a list of notable gunfounders who practised their craft in England from the fourteenth century
3 Rogers, *Artillery Through the Ages*, 27–28
4 For a detailed account of the origins and antecedents of the Schneider company see Dredge, *The Works of Messrs. Schneider and Co.*

3 The New Look

1 Hamer, *Personal Papers of Lord Rendel*, 269–70; Dougan, *The Great Gun-Maker*, 56; Scott, *Vickers*, 25
2 Dougan, *op. cit.*, 57
3 Scott, *op. cit.*, 26
4 Dougan, *op. cit.*, 59–60
5 *ibid.*, *loc. cit.*
6 *ibid.*, 83
7 *ibid.*, 77
8 *ibid.*, 78
9 Hamer, *op. cit.*, 275–76
10 For a detailed account of the history and antecedents of the firm of Krupp see Manchester, *The Arms of Krupp*
11 Dougan, *op. cit.*, 91
12 Alfred Cochrane, quoted by Dougan, *op. cit.*, 90. See also Cochrane, *The Early History of Elswick*
13 Dougan, *op. cit.*, 92

4 The Armament Kings

1 Scott, *Vickers*, 79
2 *ibid.*, 38
3 *ibid.*, 79
4 *ibid.*, 150
5 *ibid.*, 68

6 *ibid.*, 75

7 Vickers archives. On the assumption that payments sanctioned by the directors at meetings in February or March related to the previous year, the payments made for each of the twelve years from 1902 to 1913 were (to the nearest £) as follows:

	£
1902	33,667
1903	35,349
1904	40,436
1905	85,771
1906	82,677
1907	53,245
1908	39,884
1909	40,738
1910	37,216
1911	47,660
1912	64,580
1913	83,207
	£644,430

5 *The Man of Mystery: Zaharoff*

1 Neuman, *Zaharôff, The Armaments King*, deals at some length with this transaction

2 Vickers archives

3 Testimony of Mrs Orbach in Vickers archives

4 Neumann, *op. cit.*, mentions a tailoring business in southern Russia owned in the second quarter of the nineteenth century by some people called Zacharoff. The suggestion is that a family named Zacharias or Zacharios may have been one of a number of Greek families known to have fled from Constantinople after a massacre in 1822; that Zaharoff's grandfather, adopting a Russianized version of his surname, lived for a time at Kishinev, in Moldavia, and afterwards at Odessa; and that early in the 1840s he returned with his wife and two sons to Turkey. Zaharoff's father Basileos or Vasiliou Zacharoff would seem, according to this hypothesis, to have married his wife Helena after the return to Turkey, when the whole family was living in the unsavoury Tatvala quarter of Constantinople. The family then moved to Mughla, but returned about 1852 to Tatvala and lived there for the next ten years or so. Vasiliou Zacharoff is supposed to have fallen on hard times after his son's birth but to have had three more children, all girls. Two of them appeared in the photograph seen by Mrs Orbach. They are said to have emigrated to the United States but afterwards to have lived in Paris on their brother's bounty

5 Vickers archives

6 Neumann, *op. cit.*

7 *The Times* (London), January 14 and 17 and February 4, 1873

8 Scott, *Vickers*, 39; Vickers archives
9 Vickers archives
10 Scott, *op. cit.*, 41
11 Vickers archives
12 *ibid.*

6 Armageddon

1 Ritter, *The Schlieffen Plan*, gives detailed accounts of Schlieffen's and Moltke's plans
2 Tuchman, *The Guns of August*, 119
3 According to Manchester, *The Arms of Krupp*, Friedrich Krupp AG not only owned the *Benesloet*'s cargo of nickel, but also controlled the mines that produced it
4 Scott, *Vickers*, 101
5 The account given by Lloyd George in his *War Memoirs* should be compared with the official *History of the Ministry of Munitions*
6 Lloyd George, *War Memoirs*, 181

7 American Intervention and the Armistice

1 Seymour, *The Intimate Papers of Colonel House*, i, 299; Tuchman, *The Guns of August*, 338
2 Seymour, *op. cit.*, ii, 317
3 Schwarz, *American Strategy: A New Perspective*, 8
4 *ibid.*, 10
5 Lloyd George, *War Memoirs*, 1797
6 *ibid.*, 1830

8 Post-War

1 Scott, *Vickers*, 86; Vickers archives
2 Scott, *op. cit.*, 147
3 Quoted by Scott, *op. cit.*, 158
4 Vickers archives
5 Manchester, *The Arms of Krupp*, deals at some length with Krupp's response to this development
6 Scott, *op. cit.*, 84

9 Prelude to Rearmament

1 Schwarz, *American Strategy: A New Perspective*, 20
2 Manchester, *The Arms of Krupp*, 404
3 Avon, *Facing the Dictators*, 39
4 *Hearings before the Special Committee Investigating the Munitions Industry, United States Senate, 73rd Congress* (Washington, 1934–1935)
5 Scott, *Vickers*, 242
6 *ibid.*, 247–48
7 *Report of the Royal Commission on the Private Manufacture of Arms* (Cmd 5292. 1936)
8 Scott, *op. cit.*, 253

10 Rearmament

1 Jones, *Japan's New Order in East Asia*, 5
2 Scott, *Vickers*, 224–25
3 Avon, *Facing the Dictators*, 141
4 Postan, *British War Production*, 484
5 Jackson, *Air War Over France*, 20–22
6 *ibid.*, 76, 146
7 Goutard, *The Battle of France*, 39
8 *ibid.*, 27
9 Postan, *British War Production*, 6
10 *ibid.*, *loc. cit.*
11 *ibid.*, 7
12 *ibid.*, 29
13 *ibid.*, 30
14 *ibid.*, 28–29, 33–34
15 Postan, Hay and Scott, *Design and Development of Weapons*, 309–10
16 Postan, *British War Production*, 103
17 Leighton and Coakley, *Global Logistics and Strategy, 1940–1943*, 21
18 *ibid.*, *loc. cit.*
19 Hall, *North American Supply*, 119

11 World War II: The Lull and the Storm

1 Postan, *British War Production*, 103
2 Hall, *North American Supply*, 119
3 *ibid.*, 124
4 Goutard, *The Battle of France*, 131
5 Jackson, *Air War Over France*, 71, 127–30, 135
6 Hall, *op. cit.*, 128–29
7 Postan, *op. cit.*, 484
8 Hall, *op. cit.*, 131
9 *ibid.*, 138
10 *ibid.*, 141–42
11 *ibid.*, 142
12 *ibid.*, 144
13 *ibid.*, 145
14 Postan, *op. cit.*, 61

12 Arms and the 'Grand Alliance'

1 Postan, *British War Production*, 470
2 Scott, *Vickers*, 293
3 Postan, *op. cit.*, 61
4 *ibid.*, 116
5 *ibid.*, 484
6 Collier, *The Defence of the United Kingdom*, 452, 463–67
7 *ibid.*, 492
8 *ibid.*, 73
9 *ibid.*, 125
10 Hall, *North American Supply*, 428

11 Leighton and Coakley, *Global Logistics and Strategy, 1940–1943*, 28
12 Quoted by R. W. Clark, *Tizard*
13 Hall, *op. cit.*, 187
14 Leighton and Coakley, *op. cit.*, 43
15 *ibid.*, *loc. cit.*; Schwarz, *American Strategy: A New Perspective*, 31
16 Leighton and Coakley, *op. cit.*, 43
17 Hall, *op. cit.*, 428
18 Hancock and Gowing, *British War Economy*, 359
19 Leighton and Coakley, *op. cit.*, 100; Gwyer, *Grand Strategy Volume III Part I*, 151–55
20 Leighton and Coakley, *op. cit.*, 101
21 *ibid.*, 559
22 *ibid.*, *loc. cit.*
23 *ibid.*, 352 ff
24 Jones, *Japan's New Order in East Asia*, 312
25 Leighton and Coakley, *op. cit.*, 732
26 Hall, *op. cit.*, 354
27 *ibid.*, 382
28 *ibid.*, 383
29 *ibid.*, 384
30 *ibid.*, 385
31 *ibid.*, 357
32 *ibid.*, 389
33 *ibid.*, 428
34 Hancock and Gowing, *op. cit.*, 369–70
35 *ibid.*, 370
36 Postan, *op. cit.*, 389
37 *ibid.*, 484
38 *ibid.*, 471; Hall, *op. cit.*, 424
39 Postan, *op. cit.*, 424

13 World without Peace

1 Hancock and Gowing, *British War Economy*, 518–21
2 *ibid.*, 523
3 *ibid.*, 527–29
4 Thayer, *This War Business*, 17
5 Statement by British Minister of Defence in House of Commons, April 27, 1966
6 Quoted by Stanley and Pearton, *The International Trade in Arms*, 108
7 *ibid.*, 172
8 Resolution of November 20, 1965
9 Stanley and Pearton, *op. cit.*, 176

Short Bibliography

Works listed below include (a) a number of books bearing more or less directly on questions of design, manufacture, procurement, supply or cognate matters; (b) a selection, necessarily somewhat arbitrary, of books recommended for further reading or consulted for background information. Accounts of particular battles or campaigns, and in particular the 'campaign' volumes of British and American official histories, contain much valuable material; but they are so numerous that their inclusion would have made the bibliography inordinately long. With the exception of a few cited to support statements in the text, they have therefore been excluded.

OFFICIAL HISTORIES

British Official Histories: World War I

History of the Great War:

JONES, H. A. *The War in the Air*. Vols. 2–5. Oxford: Clarendon Press, 1928–35

RALEIGH, SIR WALTER. *The War in the Air*. Vol. 1. Oxford: Clarendon Press, 1922

*

History of the Ministry of Munitions, 1915–1919. 12 vols., compiled by various hands and completed in 1922. Intended primarily for official use, but the first eight volumes can be consulted in a number of academic libraries.

*

British Official Histories: World War II

History of the Second World War, United Kingdom Civil Series:

ASHWORTH, WILLIAM. *Contracts and Finance*. London: HMSO and Longmans, Green, 1953

BEHRENS, C. B. A. *Merchant Shipping and the Demands of War*. London: HMSO and Longmans, Green, 1955

GOWING, MARGARET M. (See also HANCOCK, W. K.). *Britain and Atomic Energy, 1939–1945*. London: HMSO and Longmans, Green, 1964

HALL, H. DUNCAN, and WRIGLEY, C. C. *Studies of Overseas Supply*. London: HMSO and Longmans, Green, 1956

HANCOCK, W. K., and GOWING, M. M. *British War Economy*. London: HMSO, 1949

HAY, D. See POSTAN, M. M.

HORNBY, WILLIAM. *Factories and Plant*. London: HMSO and Longmans, Green, 1958

HUGHES, RICHARD. See SCOTT, J. D.

HURSTFIELD, J. *The Control of Raw Materials*. London: HMSO and Longmans, Green, 1953

INMAN, P. *Labour in the Munitions Industries*. London: HMSO and Longmans, Green, 1957

MEDLICOTT, W. M. *The Economic Blockade*. 2 vols. London: HMSO and Longmans, Green, 1952 and 1959

O'BRIEN, TERENCE H. *Civil Defence*. London: HMSO and Longmans, Green, 1955

PARKER, H. M. D. *Manpower*. London: HMSO and Longmans, Green, 1957

POSTAN, M. M. *British War Production*. London: HMSO and Longmans, Green, 1952

POSTAN, M. M., HAY, D., and SCOTT, J. D. *Design and Development of Weapons*. London: HMSO and Longmans, Green, 1964

SAYERS, R. S. *Financial Policy, 1939–1945*. London: HMSO and Longmans, Green, 1956

SCOTT, J. D. (See also POSTAN, M. M.), and HUGHES, RICHARD. *The Administration of War Production*. London: HMSO and Longmans, Green, 1955

WRIGLEY, C. C. See HALL, H. DUNCAN

*

History of the Second World War, United Kingdom Military Series:

BUTLER, J. R. M. Grand Strategy Volume II. London: HMSO, 1957

—— Grand Strategy Volume III, Part II. London: HMSO, 1964

COLLIER, BASIL. *The Defence of the United Kingdom.* London: HMSO, 1957

EHRMAN, JOHN. *Grand Strategy Volumes V and VI.* London: HMSO, 1956

FRANKLAND, NOBLE. See WEBSTER, SIR CHARLES

GIBBS, N. H. *Grand Strategy Volume I.* London: HMSO, 1976

GWYER, J. M. A. *Grand Strategy Volume III, Part I.* London: HMSO, 1964

MARSHALL, HOWARD. *Grand Strategy Volume IV.* London: HMSO, 1972

ROSKILL, S. W. *The War at Sea.* 3 volumes in 4 parts. London: HMSO, 1956–61

WEBSTER, SIR CHARLES, and FRANKLAND, NOBLE. *The Strategy Air Offensive Against Germany.* 4 volumes. London: HMSO, 1961

*

United States Official Histories: World War II

United States Army in World War II:

COAKLEY, R. W. See LEIGHTON, R. M.

FAIRCHILD, BYRON, and GROSSMAN, JONATHAN. *The Army and Industrial Manpower.* Washington: Office of the Chief of Military History, 1959

LEIGHTON, R. M., and COAKLEY, R. W. *Global Logistics and Strategy, 1940–1943.* Washington: Office of the Chief of Military History, 1955

—— *Global Logistics and Strategy, 1943–1945.* Washington: Office of the Chief of Military History, 1968

MATLOFF, MAURICE. *Strategic Planning for Coalition Warfare, 1943–1944.* Washington: Office of the Chief of Military History, 1959

MATLOFF, MAURICE, and SNELL, E. M. *Strategic Planning for Coalition Warfare, 1941–1942.* Washington: Office of the Chief of Military History, 1953

RUPPENTHAL, ROLAND G. *Logistical Support of the Armies.* 2 vols. Washington: Office of the Chief of Military History, 1953 and 1959

SMITH, R. ELBERTON. *The Army and Economic Mobilization.* Washington: Office of the Chief of Military History, 1959

SNELL, E. M. See MATLOFF, MAURICE

WATSON, MARK S. *Chief of Staff: Pre-War Plans and Preparations.* Washington: Historical Division, Department of the Army, 1950

*

The Army Air Forces in World War II:

CRAVEN, WESLEY F., and CATE, JAMES L. (Editors). *The Army Air Forces in World War II.* 6 vols. Chicago: University of Chicago Press, 1948–1955

*

The United States Navy in World War II:

MORISON, SAMUEL ELIOT. The History of United States Naval Operations in World War II. 15 vols. Boston: Little, Brown, 1947–62

GENERAL WORKS

AVON, THE EARL OF. *The Eden Memoirs: Facing the Dictators.* London: Cassell, 1962

BALDWIN, HANSON W. *The Great Arms Race.* New York: Frederick A. Praeger, 1958

BEELER, JOHN. *Warfare in Feudal Europe.* Ithaca, N.Y.: Cornell University Press, 1971

BRET, PAUL-LOUIS. *Au Feu des Evénements.* Paris: Plon, 1959

BROCKWAY, FENNER. *The Bloody Traffic.* London: Gollancz, 1933

BULLOCK, ALAN. *Hitler: A Study in Tyranny.* London: Odhams, 1952

BURNS, J. M. *Roosevelt: The Lion and the Fox.* London: Secker and Warburg, 1957

CABLE, JAMES. *Gunboat Diplomacy.* London: Chatto and Windus for the Institute for Strategic Studies, 1971

CALDER, ANGUS. *The People's War: Britain, 1939–1945.* London: Cape, 1969

CARR, J. C., and TAPLIN, W. *History of the British Steel Industry.* Oxford: Basil Blackwell, 1962

CHANDOS, VISCOUNT. *The Memoirs of Lord Chandos*. London: The Bodley Head, 1962

CHURCHILL, WINSTON S. *The Second World War*. 6 vols. London: Cassell, 1948–54

CLARK, R. W. *The Birth of the Bomb*. London: Phoenix House, 1961

CLAY, SIR HENRY. *Lord Norman*. London: Macmillan, 1957

COCHRANE, ALFRED. *The Early History of Elswick*. Newcastle-upon-Tyne: Mawson Swann and Morgan, 1909

CROSBY, J. R. *Disarmament and Peace in British Politics, 1914–19*. Harvard: Harvard University Press, 1957

DEMPSTER, DEREK. See WOOD, DEREK

DOUGAN, DAVID. *The Great Gun-Maker*. Newcastle-upon-Tyne: Frank Graham, no date (1971)

DREDGE, J. The Works of Messrs. Schneider and Company. London: The Bedford Press, 1900

ENGELBRECHT, H. C., and HANIGHEN, F. C. *Merchants of Death*. New York: Dodd, Mead, 1934

FEIS, H. *The Road to Pearl Harbor*. Princeton: University Press, 1950

—— *Churchill, Roosevelt, Stalin*. Princeton: University Press, 1957

—— *The China Tangle*. New York: Atheneum, 1965

FISCHER, FRITZ. *Germany's Aims in the First World War*. London: Chatto and Windus, 1967

FORDE, C. D. *Habitat, Economy and Society*. London: Methuen, 1933

GEORGE, DAVID LLOYD. *War Memoirs*. 2-volume edition. London: Odhams, 1936

GIBBS-SMITH, CHARLES H. *The Aeroplane: An Historical Survey of its Origins and Development*. London: Science Museum, 1960

—— *Aviation: An Historical Survey from its Origins to the End of World War II*. London: HMSO, 1970

GOUTARD, COLONEL A. *The Battle of France, 1940*. London: Frederick Muller, 1958

GRANT, SIR ALLAN. *Steel and Ships: The History of John Brown's*. London: Michael Joseph, 1950

GROUEFF, STÉPHANE. *Manhattan Project*. London: Collins, 1967

GROVES, LIEUTENANT-GENERAL LESLIE R. *Now It Can be Told*. London: André Deutsch, 1963

HAMER, F. E. (Editor). *The Personal Papers of Lord Rendel*. London: Ernest Benn, 1931

HANIGHEN, F. C. See ENGELBRECHT, H. C.

HANKEY, LORD. *The Supreme Command, 1914–1918*. London: George Allen and Unwin, 1961

HELMER, WILLIAM J. *The Gun that Made the Twenties Roar*. London: The Macmillan Company/Collier-Macmillan Ltd, 1969

HOGG, BRIGADIER OLIVER F. G. *Artillery: Its Origin, Heyday and Decline*. London: C. Hurst and Company, 1970

JACKSON, ROBERT. *Air War over France, May–June, 1940*. Shepperton: I. Allan, 1974

JAQUES, LIEUTENANT W. H., United States Navy. *The Establishment of Steel Gun Factories in the United States*. Annapolis: U.S. Naval Institute, 1884

JONES, F. C. *Japan's New Order in East Asia*. London: Oxford University Press, 1954

LIVINGSTON, BRIGADIER-GENERAL GUY. *Hot Air in Cold Blood*. London: Selwyn and Blount, 1933

LYTTELTON, OLIVER. See CHANDOS, VISCOUNT

MACLEOD, IAIN. *Neville Chamberlain*. London: Frederick Muller, 1961

MANCHESTER, WILLIAM. *The Arms of Krupp, 1857–1968*. London: Michael Joseph, 1969

MARDER, ARTHUR J. *British Naval Policy, 1880–1905*. London: Putnam, 1941

MAXIM, H. *My Life*. London: Methuen, 1915

MORISON, SAMUEL ELIOT. *The Two-Ocean War*. Boston: Little, Brown, 1963

NEUMANN, ROBERT. *Zaharoff, The Armaments King*. London: George Allen and Unwin, 1938

NOEL-BAKER, PHILIP. *The Private Manufacture of Armaments*. 2 vols. London: Gollancz, 1936

PEARTON, MAURICE. See STANLEY, JOHN

PIGOU, A. C. *Aspects of British Economic History, 1918–25*. London: Macmillan, 1947

PRATT, JULIUS W. *Cordell Hull*. New York: Cooper Square Publishers, 1964

RENDEL, STUART. See HAMER, F. E.

RITTER, GERHARD. *The Schlieffen Plan.* London: Oswald Wolff, 1958

ROGERS, COLONEL HUGH C. B. *Artillery Through the Ages.* London: Seeley Service, 1971

SAMPSON, ANTHONY. *The Arms Bazaar.* London: Hodder and Stoughton, 1977

SCHWARZ, URS. *American Strategy: A New Perspective.* New York: Doubleday, 1966

SCOTT, J. D. *Vickers: A History.* London: Weidenfeld and Nicolson, 1962

SCRIVENOR, HARRY. *A Comprehensive History of the Iron Trade Throughout the World.* London: Smith, Elder, 1841

SHERWOOD, ROBERT E. *The White House Papers of Harry L. Hopkins.* 2 vols. London: Eyre and Spottiswoode, 1948–49

STANLEY, JOHN, and PEARTON, MAURICE. *The International Trade in Arms.* London: Chatto and Windus for the International Institute for Strategic Studies, 1972

TAPLIN, W. See CARR, J. C.

TENNENT, SIR J. EMERSON. *The Story of the Guns.* London: Longmans, Green, 1864

TUCHMAN, BARBARA W. *The Guns of August.* New York: The Macmillan Company, 1962

WARLIMONT, WALTER. *Inside Hitler's Headquarters, 1939–1945.* London: Weidenfeld and Nicolson, 1964

WELLES, SUMNER. *The Time for Decision.* New York: Harper and Row, 1944

WERTH, ALEXANDER. *Russia at War, 1941–1945.* London: Barrie and Rockliff, 1964

WOOD, DEREK, and DEMPSTER, DEREK. *The Narrow Margin.* London: Hutchinson, 1961

WOODWARD, E. L. *Great Britain and the German Army.* London: Oxford University Press, 1935

YOUNG, G. M. *Stanley Baldwin.* London: Rupert Hart-Davis, 1952

Index